GIS Tutorial for Health Professionals

Esra Ozdenerol, PhD

INGI, LLC

Copyright @ 2025 INGI, LLC
All rights reserved.

PREFACE

This book, GIS Tutorial for Health Professionals, introduces GIS and its applications to public health through hands-on learning. It covers fundamental topics like spatial analysis, including John Snow's historic cholera outbreak mapping, and explores GIS applications in chronic diseases, birth health outcomes, infectious diseases, and children's lead poisoning. The book provides practical exercises designed to teach GIS skills, making it accessible for health professionals at all levels. The book also covers spatial statistics in GIS, including spatial regression techniques and hotspot maps, to assist health professionals in analyzing and interpreting complex health data.

The book has seven chapters. Each chapter is divided into two sections: *Mastering Concepts* and *Mastering Skills*. In the *Mastering Concepts* section, GIS applications to a specific health topic are explored in detail, providing readers with a thorough understanding of how GIS can address public health challenges. The *Mastering Skills* section follows with GIS tutorial exercises related to the topic covered in the *Mastering Concepts* section, allowing readers to apply GIS techniques to that heath topic. Each exercise has a series of steps, input datasets, instructions on the structure required for GIS analysis, and maps for you to produce.

Additionally, it highlights the use of online GIS tools such as dashboards and story maps, which empower non-GIS users to effectively analyze and visualize health data. These tools enhance the ability to engage with and interpret complex datasets, supporting better decision-making and improving public health outcomes.

The book is designed for a diverse audience, including health professionals, healthcare managers, practitioners, public health and health management students, computer specialists aspiring to work in the health sector, as well as academics and researchers seeking to enhance their expertise in GIS.

This book is primarily designed as a GIS tutorial and computer lab textbook, but it can also be used for self-study. It is suitable for an entire semester or can be adapted for shorter courses, with selected chapters for two to three-day sessions. In working through this book, the following sequence of steps is suggested: Read through the Mastering the Concepts section of each chapter to get familiar with the health topic and work on the Mastering the Skills section for a step-by-step tutorial and explanation of key techniques.

If you are new to Arc GIS Pro and are using the book as a self-study guide, we recommend you work through the chapters in sequence. However, because the chapters are largely independent of each other, you can use them in the order that best fits your needs. The book also comes with a course platform that readers can purchase, providing step-by-step video instructions, lectures, and answer keys to reinforce learning.

I would like to express my gratitude to the many individuals who have used and provided feedback on this tutorial. I hope it continues to serve their needs in the ever-evolving field of GIS in health.

I also hope this tutorial has highlighted the tremendous potential of GIS as a valuable tool for health professionals. Although it focuses on a few key health applications, the possibilities for GIS to contribute to a wide range of health areas are vast. The techniques and examples shared are intended to guide health professionals in utilizing GIS to enhance research, inform decision-making, and address pressing global health challenges, making it a valuable resource for both U.S. and international contexts.

NOTICE: Arc GIS Pro® desktop, Arc GIS StoryMaps, ArcGIS Experience Builder used in this text are registered trademarks of ESRI, Inc. The software names and the screen shots used in the text are reproduced by permission. For ease of reading only, the ® and ™ symbols have been omitted from the names; however, no infringement or denial of the rights of ESRI® is thereby intended or condoned by the author.

Certain Esri Imagery in this work are owned by Esri and its data contributors and are used herein with permission. Copyright © 2024 Esri and its data contributors. All rights reserved.

PREVIOUS EXPERIENCE

This book assumes that the reader is comfortable using Windows™ to carry out basic tasks such as copying files, moving directories, opening documents, exploring folders, and editing text and word processing documents. Previous experience with maps and map data is also helpful. No previous GIS experience or training is necessary to use this book.

SYSTEM REQUIREMENTS

ESRI has updated the minimum system requirements for ArcGIS Pro on their official website, www.esri.com. The system requirements for ArcGIS Pro 3.4 provide a comprehensive overview of both hardware and software specifications necessary for optimal performance. These specifications are designed to support a wide range of workflows, from basic GIS tasks to advanced spatial analysis. For detailed and up-to-date information, please visit ESRI's official page at www.esri.com.

Minimum System Requirements:

Operating System: Windows 10 (64-bit) or Windows 11 (64-bit)

Processor: Quad-core 2.2 GHz processor or faster

Memory: 8 GB of RAM

Storage: 32 GB of available hard-drive space

Graphics: DirectX 11 compatible graphics card with at least 4 GB of dedicated VRAM

Display: Minimum 1024x768 resolution (1080p or higher recommended)

Internet: Internet connection for licensing and updates

Microsoft .NET: Desktop Runtime 8.0.0 or later

Recommended System Requirements:

Processor: 3.0 GHz or faster with multiple cores

Memory: 16 GB of RAM or more

Graphics: NVIDIA GPU with at least 6 GB VRAM, CUDA compute capability 6.1 or later

Storage: SSD recommended for faster read/write speeds

Additional Software: WebView2 Runtime version 117 or later

Software:

Windows 10 (64-bit) or Windows 11 (64-bit)

A web browser, such as Google Chrome, Mozilla Firefox, or Microsoft Edge

Arc GIS Pro 3.4 or higher; ESRI apps such as Arc GIS Story Maps, ArcGIS Experience Builder required for some exercises.

For assistance in acquiring or installing these components, contact your system administrator, hardware/software provider, or local computer store.

Resources Page

All supporting materials for this tutorial are available on the **Resources** page. Navigate to the Resources page at www.esraozdenerol.com/resources
There you'll find:

- **Tutorial Data Files** – Download all datasets used throughout the lessons so you can follow along and practice each step hands-on.

- **Companion Course (with Answer Key Included)** – Gain access to the full course, which includes lecture presentations explaining each concept, plus step-by-step video walkthroughs of the exercises. The complete answer key is included as part of the course package, so you can check your work and deepen your understanding.

Visit the Resources page anytime to download materials or purchase the full course for expanded content and guided practice.

Tutorial Data:

You will need the tutorial data and access to Arc GIS Pro ® desktop GIS software and ESRI apps such as Arc GIS StoryMaps and ArcGIS Experience Builder to perform the exercises in this book. Access to Microsoft Word, Excel, and Internet connection are also necessary for some exercises.

All the datasets and materials needed to follow along with this tutorial can be found on the Resources page of the author's official website. The Resources page provides direct access to the tutorial's GitHub repository, where all datasets, project files, and supplementary materials are securely hosted.

Recommended Download Procedure

Navigate to the Resources page at www.esraozdenerol.com/resources

Follow the provided link to the GitHub repository.

On the GitHub page, click the green "Code" button.

Select "Download ZIP" to obtain a compressed archive of the complete repository.

Extract the ZIP file to access all chapter exercise data folders, the main project file, and accompanying documentation.

It is recommended to download the entire repository as a ZIP file rather than individual files or folders. This ensures that all files, datasets, and project settings remain properly linked and function seamlessly throughout the tutorial.

Contents of the Download Package

Chapter Exercise Data Folders – There is one folder for each chapter exercise. Each folder includes the relevant datasets, step-by-step demonstration videos, and a Word document corresponding to the exercise in the *Mastering the Skills* section of the chapter.

Main Project File – Each exercise comes with a pre-set GIS project file that integrates the necessary data (such as geodatabases) to simplify workflow management.

Documentation (Optional) – Supplementary ReadMe files or detailed instructions providing additional guidance for specific chapters, where applicable.

GIS Course for Health Professionals

Visit the author's official website and go to the www.esraozdenerol.com/resources to access and purchase a course that complements the GIS tutorial for health professionals. The course includes:

- Step-by-step video instructions
- Lecture presentations for each chapter
- Answer keys for all exercises

For instructors: You can use the lecture presentations and answer keys to support teaching and class activities.

This course is designed to guide you through the tutorial in a structured way, making it easier to learn and apply GIS concepts in health-related projects.

ABOUT THE AUTHOR

Dr. Esra Ozdenerol is the Dunavant University Professor of Geographic Information Science at the University of Memphis, where she also directs the Graduate Certificate in Geographic Information Systems (GIS). She is an Adjunct Professor of Preventive Medicine at the University of Tennessee Health Science Center and the Director of the Spatial Analysis and Geographic Education (SAGE) Laboratory at the University of Memphis.

Dr. Ozdenerol specializes in the use of GIS to address public health issues, including health disparities, infectious disease epidemiology, and environmental health. She is the author of several influential books, including Spatial Health Inequalities: Adapting GIS Tools and Data Analysis, Gender Inequalities: GIS Approaches to Gender Analysis, and The Role of GIS in COVID-19 Management and Control. She has been recognized by organizations such as AAG for her work in GIS and public health.

In addition to her academic work, Dr. Ozdenerol owns a GIS consulting company specializing in GIS applications for health and providing training for health professionals. She has lectured and led workshops on GIS applications in health both nationally and internationally.

You can contact Dr. Ozdenerol through her website at www.esraozdenerol.com.

ACKNOWLEDGEMENTS

This tutorial would not have been possible without the support, dedication, and contributions of many individuals to whom I am deeply grateful.

First and foremost, I extend my heartfelt thanks to my assistant, Mekensie Ivy, for her meticulous work in formatting this tutorial. Her attention to detail and commitment ensured that the final product met the highest standards.

I am also deeply appreciative of Joana Goebel, whose artistry and vision brought this tutorial to life with a fantastic cover design. Her creativity perfectly encapsulates the spirit of this work.

A special note of gratitude goes to my former students for their invaluable assistance and feedback with the exercises. Their expertise and enthusiasm were instrumental in shaping the practical aspects of this tutorial, including the development of the app exercises. Special thanks to my PhD student, Rebecca Bingham-Byrne, for her valuable assistance in compiling several exercises for this tutorial from my years of accumulated materials, former classes, and curated collections. I would also like to recognize my former GIS certificate student, David L. Fawcett, an independent GIS and remote sensing researcher from Spring Hill, TN, for his contribution to Exercise 7 on building a simple ArcGIS Experience Builder.

ESRI, Inc., was prompt and generous in its granting of permission to use the screen shots data and other materials throughout the text. I extend heartfelt thanks to Centers for Disease Control (CDC), U.S. Census Bureau, City of London, National Historical Geographic Information System (NHGIS) at the University of Minnesota, State of Tennessee, for putting their fine GIS and health data sets in the public domain.

To my students over the years, thank you for your curiosity, dedication, and passion for learning. You have continuously inspired me to refine my teaching and expand my perspective. Your growth and achievements have been among my greatest joys as an educator.

To my family, my two sons, Derin and Deniz, thank you for being my anchors. Your resilience inspires me daily, and I strive to be strong for you. To my extended family across the ocean—my parents, my sister, my nieces, and my brother-in-law—your unwavering support and kindness have been a source of strength and encouragement throughout this journey. To my dear friend Funda, your steadfast friendship has been a pillar of support, and I am forever grateful.

This tutorial is a reflection of the collective effort, support, and inspiration I have received from each of you. Thank you for helping me bring this vision to fruition.

ACRONYMS

ACS - American Community Survey

BLLs - Blood Lead Levels

BMI - Body Mass Index

CDC - Centers for Disease Control and Prevention

COVID-19 - Coronavirus Disease 2019

EBLLs - Elevated Blood Lead Levels

Ebola - Ebola Virus Disease

GIS - Geographic Information Systems

HIV/AIDS - Human Immunodeficiency Virus / Acquired Immunodeficiency Syndrome

LD - Lyme Disease

LBW - Low Birth Weight

LSES - Low Socioeconomic Status

NASA - National Aeronautics and Space Administration

NAICS - North American Industry Classification System

NHGIS - National Historical Geographic Information System

NOAA - National Oceanic and Atmospheric Administration

NCI - National Cancer Institute

RS - Remote Sensing

SES - Socioeconomic Status

STDs - Sexually Transmitted Diseases

U.S. - United States

USGS - United States Geological Survey

Table of Contents

PREFACE 3

ACKNOWLEDGEMENTS 9

ACRONYMS 10

CHAPTER 1: THE ROLE OF GIS IN PUBLIC HEALTH 14

Mastering the Concepts 14

Mastering the Skills 16

Exercise 1: Introduction to ArcGIS Pro – Setting up for success in ArcGIS, exploring data models and data types, and working in the ArcGIS environment 16

    Section 1.1: Create and access an ArcGIS Online profile 17

    Section 1.2: Download and open ArcGIS Pro on your computer 19

Section 2: Add basemaps, vector data, and raster data to your mapping project 22

    Section 2.1: Adding basemaps 22

    Section 2.2: Adding vector data 24

    Section 2.3: Adding raster data 29

Section 3: Working in the ArcGIS Environment 31

    Section 3.1: View and edit metadata for layers in ArcGIS Pro 31

    Section 3.2: Select certain features from a layer 36

    Section 3.3: Export spatial data and maps 40

CHAPTER 2: SPATIAL ANALYSIS AND HEALTH 45

Mastering the Concepts 45

Mastering the Skills 48

Exercise 2: Visualizations in ArcGIS Pro – Learning basic tools and how to visualize spatial data . 48

    Section 1.1: Open an ArcGIS Pro project 49

    Section 1.2: Set Data Source 50

    Section 1.3: Spatial join 53

    Section 1.4: Creating a Buffer 55

    Section 1.5: Changing Symbol Graphic Elements 57

Section 2: How to Visualize Spatial Data 61

Section 2.1: Graduated Symbols 61

Section 2.2: Choropleth maps 64

Section 2.3: Creating a heat map 66

CHAPTER 3. CHRONIC DISEASES 68

Mastering the Concepts 68

Mastering the Skills 74

Exercise 3: Using Health Data in ArcGIS Pro – Data cleaning, geocoding, and health data visualization 74

Section 1.1: Clean up excel files with health information and upload into ArcGIS Pro 76

Section 2: Geocoding 81

Section 2.1: Conduct address matching to geographic coordinate analysis (geocoding) 81

Section 2.2: Adding relates between shapefiles and excel files 92

Section 3.1: Create meaningful maps 99

CHAPTER 4: BIRTH HEALTH 118

Mastering the Concepts 118

Mastering the Skills 121

Exercise 4: Birth Health Outcomes – Creating a Health Story using ArcGIS StoryMaps 121

Section 1.1: Use health data to create an interesting story 122

CHAPTER 5: INFECTIOUS DISEASES 129

Mastering the Concepts 129

Mastering the Skills 132

Exercise 5: Spatial statistics of Health Data in ArcGIS Pro – Clustering of values to search for spatial patterns 132

Section 1.1: Calculate spatial autocorrelation using Getis-Ord General G 133

Section 1.2: Calculate spatial autocorrelation using Moran's I 136

Section 2: Local Level 139

Section 2.1: Calculate spatial autocorrelation using Anselin Local Moran's I 139

Section 2.2: Calculate spatial autocorrelation using Getis-Ord Gi* 141

CHAPTER 6. CHILDREN'S LEAD POISONING 145

Mastering the Concepts 145

Mastering the Skills 147

Exercise 6. Spatial statistics of Health Data in ArcGIS Pro – Correlation of values to understand how one variable can impact another. 147

    Section 1.1: Running the Ordinary Least Squares Regression (OLS) tool 148

    Section 1.2: Perform six checks to create a proper model 154

Section 2: Multivariate Regression model 160

    Section 2.1: Running the Ordinary Least Squares Regression (OLS) tool 160

    Section 2.2: Perform six checks to create a proper model 163

CHAPTER 7: ONLINE GIS APPLICATIONS 167

Mastering the Concepts 167

Mastering the Skills 170

Exercise 7. Constructing a Simple ArcGIS Experience Builder Site and Configuring Multiple Widget Types. 170

    Section 1.1: Create the Experience Builder File 172

    Section 1.2: Format the Page 172

    Section 1.3: Image and Text Widgets 174

    Section 2.1: The Map and Search Widgets 176

    Section 2.2: The Text Widget and Dynamic Text 179

    Section 2.3: The Chart Widget 181

    Section 3.1: Create a New Page 183

    Section 3.2: Add a Button Widget for Navigation and Publish 183

ENGAGE WITH DR. ESRA OZDENEROL 185

CHAPTER 1: THE ROLE OF GIS IN PUBLIC HEALTH

Mastering the Concepts

This chapter explores the transformative power of Geographic Information Systems (GIS) in addressing critical challenges in public health. Geographic Information Systems (GIS) are powerful tools for analyzing and visualizing spatial data, enabling professionals in health-related fields to make informed decisions based on location-based insights. In health applications, GIS is used to map and analyze the spatial distribution of diseases, identify disease hotspots, and understand environmental risk factors. By overlaying health data with geographic information, GIS allows for better resource allocation, targeted interventions, and more effective disease prevention strategies. These capabilities help public health officials and healthcare providers address pressing health challenges and improve outcomes across diverse regions.

One of the most prominent uses of GIS in health is tracking the spread of infectious diseases. By mapping the locations of disease cases such as COVID-19, Zika, or Ebola, GIS allows health officials to identify clusters of infection and predict potential hotspots. This spatial understanding is essential for managing outbreaks and implementing effective containment measures. GIS tools enable real-time updates, which are crucial for quickly assessing the spread of diseases and optimizing intervention strategies. Similarly, GIS plays a significant role in environmental health analysis by helping identify areas with high exposure to environmental hazards, such as air pollution or contaminated water, which can contribute to a range of health issues. Mapping these environmental risks helps develop targeted public health interventions to mitigate negative health impacts in affected areas.

Health disparities often exist across different geographic regions, with certain communities facing higher rates of disease or poorer access to healthcare services. GIS is an essential tool for identifying these disparities, enabling researchers to pinpoint areas with limited access to healthcare or poor health outcomes. By visualizing these inequities, GIS helps direct resources and interventions to underserved communities, promoting health equity and ensuring that all populations receive the care they need. Furthermore, in healthcare planning, GIS is instrumental in determining the optimal locations for new facilities. By analyzing factors such as population density, access to transportation, and existing healthcare services, GIS helps planners identify areas with high demand for healthcare but insufficient supply. This guidance enables more informed decisions about where to build new hospitals, clinics, or other healthcare resources, ensuring more equitable access to care.

GIS is also widely used to assess the health needs of specific populations. By combining demographic data with health indicators, such as rates of chronic disease or vaccination coverage, GIS helps identify vulnerable groups and determine their specific health needs. This spatial approach allows public health officials to allocate resources more efficiently and target interventions to the populations most in need. Another key application of GIS in health is mapping the distribution of risk factors. Factors such as obesity rates, smoking prevalence, or socioeconomic status can vary greatly across geographic areas. By visualizing these risk factors, GIS helps identify high-risk areas and enables targeted public health initiatives aimed at reducing risk behaviors and improving overall health.

At the core of GIS in health applications is the integration of various types of data. Health data, such as disease cases, mortality rates, and demographic information, are combined with geographic data, which can include census boundaries, addresses, and environmental factors. This integration allows for a comprehensive understanding of health patterns within specific geographic regions. GIS tools are then used to perform spatial analysis, identifying patterns, clusters, and trends in health data based on geographic location. For example, by performing spatial analysis, health professionals can detect disease clusters, assess the proximity of healthcare facilities to high-risk populations, or identify environmental hazards in specific areas.

One of the most powerful features of GIS is its ability to create visual representations of data. Maps, charts, and other visual tools make it easier to communicate complex health data in an accessible way. By visualizing the distribution of health issues, GIS enables decision-makers to quickly identify areas of concern and plan appropriate interventions. An example of this in action can be seen in the mapping of cancer incidence rates. GIS is often used to identify areas with higher-than-average rates of specific cancers, helping public health officials investigate potential environmental causes or inform targeted prevention strategies, such as promoting screening in high-risk areas. Similarly, for diseases like malaria, Zika, and dengue, GIS is used to track mosquito breeding sites, helping health officials target vector control efforts, such as spraying insecticides or eliminating breeding grounds, to reduce the spread of these diseases.

GIS technology has evolved far beyond static maps, transforming into dynamic, interactive applications that empower users to engage with data in new and impactful ways. Through simple, intuitive dashboards and mobile applications, GIS enables non-specialized users to analyze geographic data effectively. These tools democratize access to complex spatial insights, allowing a broader audience to make informed, data-driven decisions.

GIS is no longer just about maps—it's about enabling action through apps that make spatial intelligence accessible, impactful, and essential. In health applications, this transformation is particularly groundbreaking. GIS dashboards and mobile apps make tracking health threats and monitoring public health data accessible in real time. The integration of geospatial tools with cloud-based analysis enables health professionals, policymakers, and even the public to visualize and respond to crises with agility. This digital transformation ensures that valuable data and insights are always at your fingertips, fostering rapid, effective responses to both global health challenges and everyday scenarios.

Finally, GIS plays a crucial role in evaluating access to healthcare. By mapping the locations of healthcare facilities and the distance to these facilities for different populations, GIS can identify underserved areas that may require additional resources, such as new clinics or mobile health services. This capability not only aids in resource allocation but also ensures that underserved populations receive timely and adequate healthcare. Through these various applications, GIS provides essential tools for improving public health, helping health professionals make data-driven decisions that ultimately create healthier communities

Mastering the Skills

Exercise 1: Introduction to ArcGIS Pro – Setting up for success in ArcGIS, exploring data models and data types, and working in the ArcGIS environment

In this tutorial you are going to become familiar with ArcGIS online and ArcGIS pro. You will set up an online account within the university's organization, download and set up ArcGIS pro on your computer, as well as get some beginning experience using the program. You will learn how to add basemaps, import spatial data, explore data models (vector and raster), visualize different data types (point, line, and polygon), edit metadata, and export spatial data.

OBJECTIVES

- Create and access an ArcGIS Online profile
- Download and open ArcGIS Pro on your computer
- Add basemaps, vector data, and raster data to your mapping project
- View and edit metadata for layers in ArcGIS Pro
- Select certain features from a layer
- Export spatial data and maps
- Save map projects

Required data:

The original data sources listed below are provided for reference only. You do not need to download or curate these datasets from the original sources. All data, preconfigured GIS project files, and geodatabases required for this exercise are already included with the tutorial data and are available through the Resources page of the author's official website: www.esraozdenerl/resources. Instructions for accessing the tutorial data can be found in the Tutorial Data section of the Preface.

Original data sources:

1. Data from the Introduction to ArcGIS Pro folder. Sources:
    a. Cholera deaths and water pumps (Point data): https://kuscholarworks.ku.edu/handle/1808/10772
    b. Road and place boundary shapefiles for the Greater London Area (Line and polygon data): http://download.geofabrik.de/europe/great-britain/england/greater-london.html
    c. Polling districts for the Greater London Area (Polygon data): https://osdatahub.os.uk/downloads/open/BoundaryLine
    d. John Snow's Original Map Georeferenced (Raster data): https://github.com/mapninja/ArcGIS-Pro-101/tree/master

Section 1: Setting up for Success in ArcGIS

Section 1.1: Create and access an ArcGIS Online profile

One of the first, and most important, things to consider when working with ArcGIS is to have an ArcGIS Online account to manage and share their information. You can either have an individual online account or be a member of a bigger organization. In some cases, the organization will create an account for you instead of inviting you to join their organization. However, sometimes you will have to create your own ArcGIS Online profile and convert it to an organizational member after being invited. Be sure to ask how your organization works before creating an ArcGIS Online profile of your own.

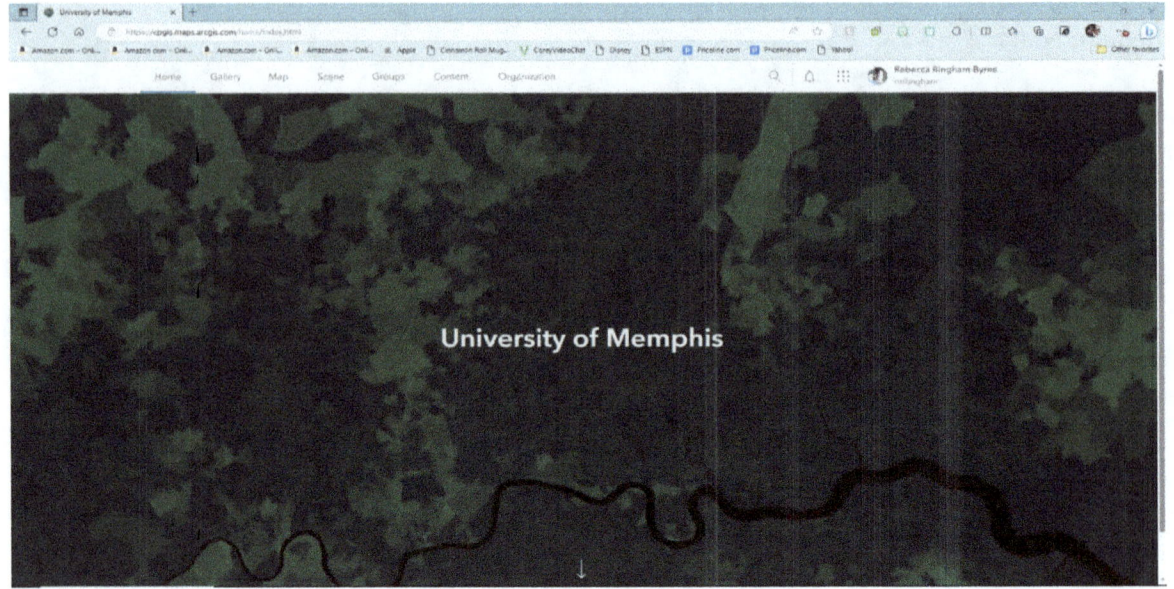

For the first part of this tutorial, you will learn how to create and access an ArcGIS Online profile.

1. Before creating your ArcGIS account, make sure to have the following information available:
    a. Your name
    b. Your email address (best to use the one provided by your university/organization)
    c. A username (can be your email address)
    d. A password with at least 8 characters. It must include at least 1 letter and 1 number. It is case sensitive and cannot be the same as your username.

    *Note: you can create a public account using a social login as well, but it may be more confusing when you enter an organization later.

2. There are two ways to create an ArcGIS Online account:
    a. You can create an individual public account by following the guidelines on this page: https://doc.arcgis.com/en/arcgis-online/get-started/create-account.htm.
    b. If you have been invited to join an organization, you can join the organization by following the guidelines on this page: https://doc.arcgis.com/en/arcgis-online/get-started/join-org.htm#ESRI_SECTION1_73A22DB113B7429B91101EA4D8A8E835

3. Once your account has been established, you can access your ArcGIS Online account by signing in with your credentials on this website: https://www.arcgis.com/index.html

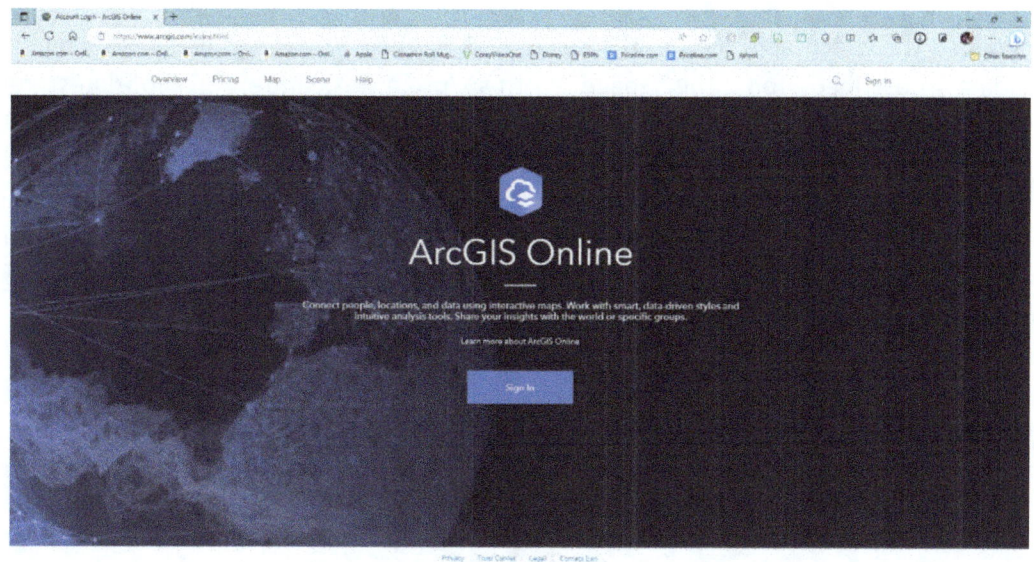

Section 1.1 Task: Submit a screenshot of your ArcGIS Online homepage.

Section 1.2: Download and open ArcGIS Pro on your computer

The ArcGIS Online account credentials are also how you sign into ArcGIS Pro. Before you can use ArcGIS Pro, you will have to install the program onto your computer. To ensure successful operations of the program, please check that your computer meets the minimum specifications outlined on this page: https://pro.arcgis.com/en/pro-app/latest/get-started/arcgis-pro-system-requirements.htm.

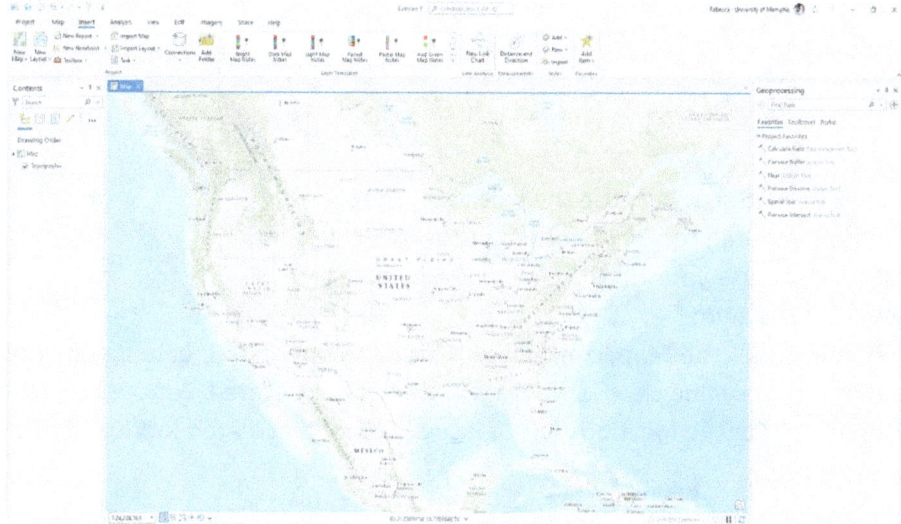

The next section of the tutorial will guide you through downloading and opening ArcGIS Pro on your computer.

1. Download the executable file (.exe) from ArcGIS Online using the guidelines under the heading described on this page: https://pro.arcgis.com/en/pro-app/latest/get-started/download-arcgis-pro.htm

2. After the executable file (.exe) has been downloaded onto your computer, run the file to install the ArcGIS Pro software onto your machine using the instructions found on this page: https://pro.arcgis.com/en/pro-app/latest/get-started/install-and-sign-in-to-arcgis-pro.htm

3. Once installed, the program will now be able to be opened, and it will look similar to the figure below.

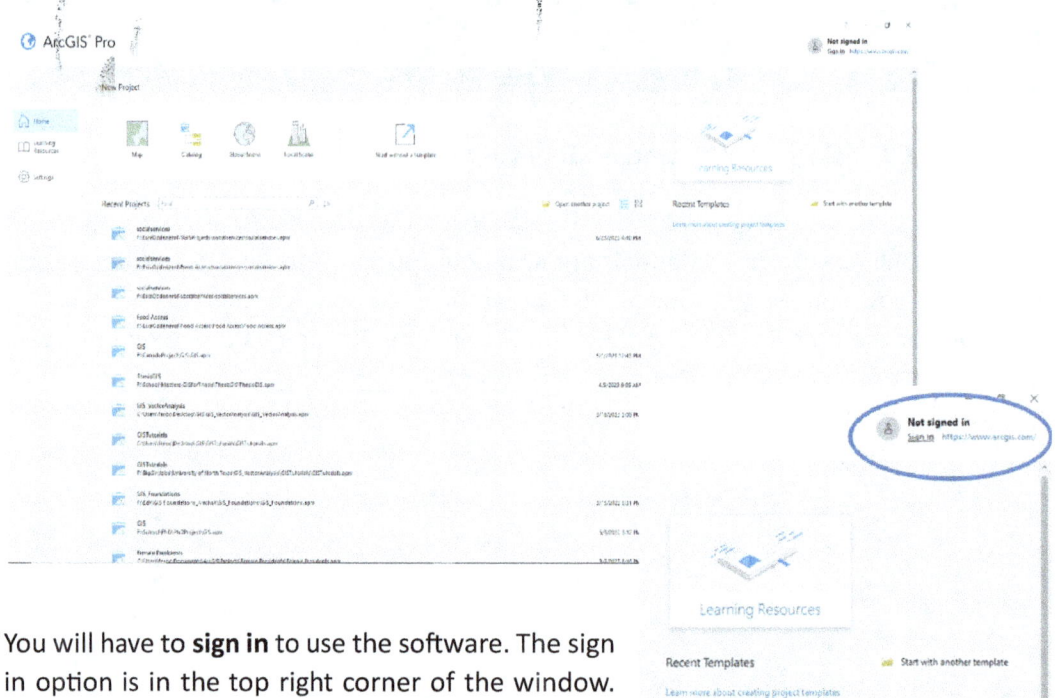

4. You will have to **sign in** to use the software. The sign in option is in the top right corner of the window. Make sure to use your ArcGIS Online credentials to sign into ArcGIS Pro.
5. Next, if you click the **Map** icon, you will have the option to create a new project. You can name the project and choose where you want it to be stored. For the **Project Name**, enter "Exercise 1". For the **location**, save it into a folder you created for the course.

20

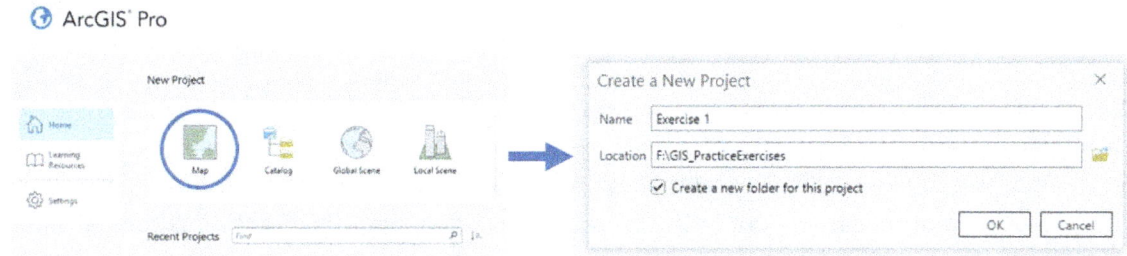

6. Once you have finished this, the software will open a new blank map.

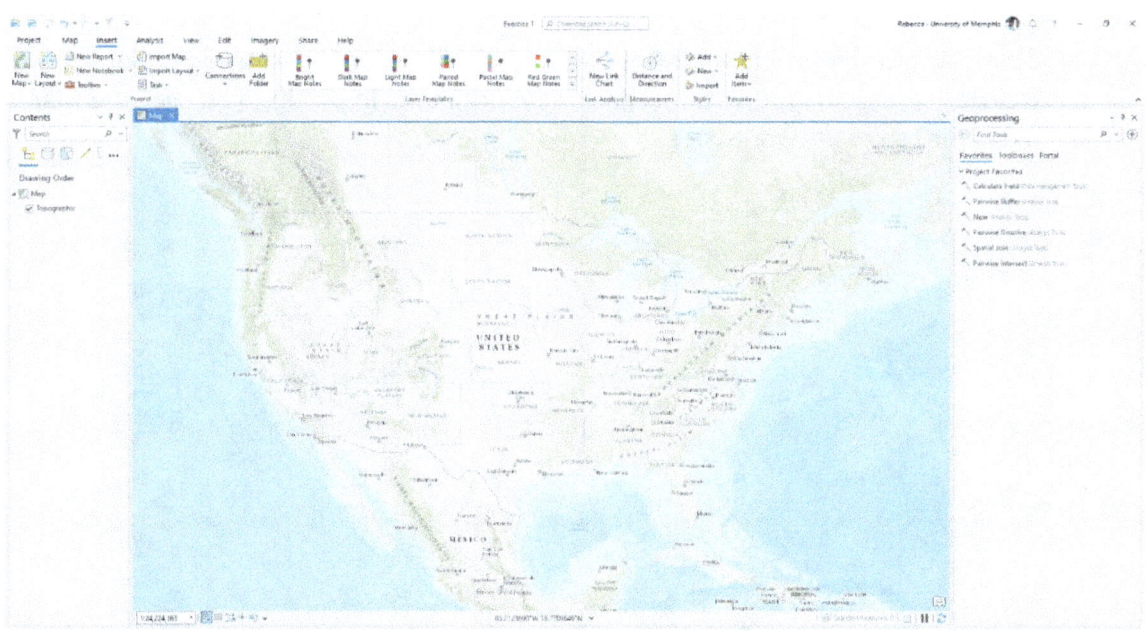

Section 1.2 Task: Submit a screenshot of your new blank map on ArcGIS Pro.

Section 2: Add basemaps, vector data, and raster data to your mapping project

Section 2.1: Adding basemaps

It is important to have a solid foundation on which users can view the location you are studying. Basemaps help users easily identify the area of focus of a map. ArcGIS provides several standard basemaps that are easily recognizable, which allows for a quick and convenient start to any mapping project.

This part of the tutorial will demonstrate how to add basemaps to your mapping project.

1. Continuing in the new blank map, you can see a basemap is already in place. You can change the basemap to better align with your needs. To do this, you click the **Map** tab along the top of the workspace.

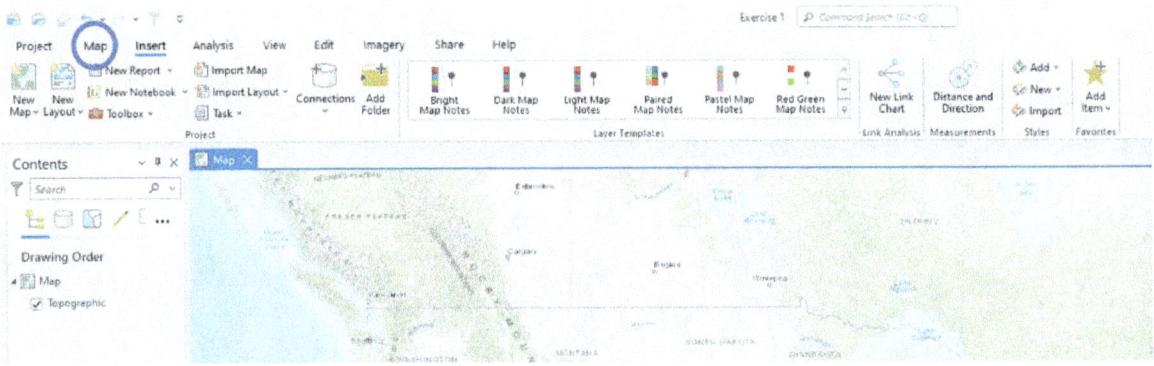

2. Click the **Basemap** icon and look at the variety of basemaps provided to you in ArcGIS Pro.

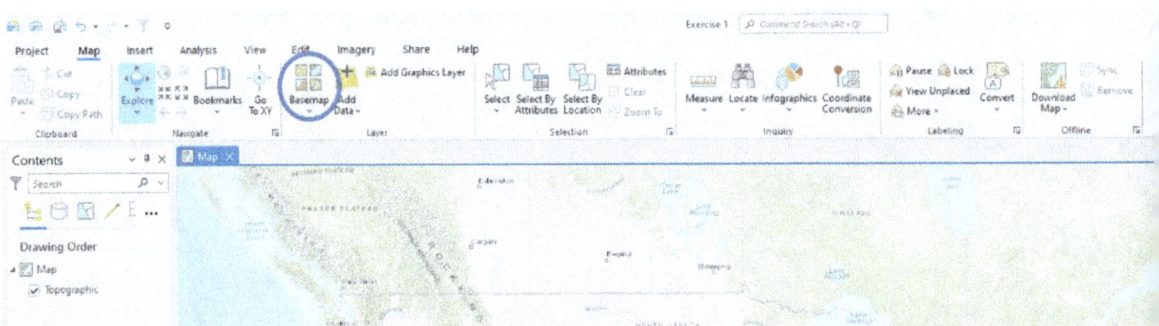

3. Choose the **OpenStreetMap** basemap from the list. **Note: the contents panel will now show OpenStreetMap instead of Topographic.**

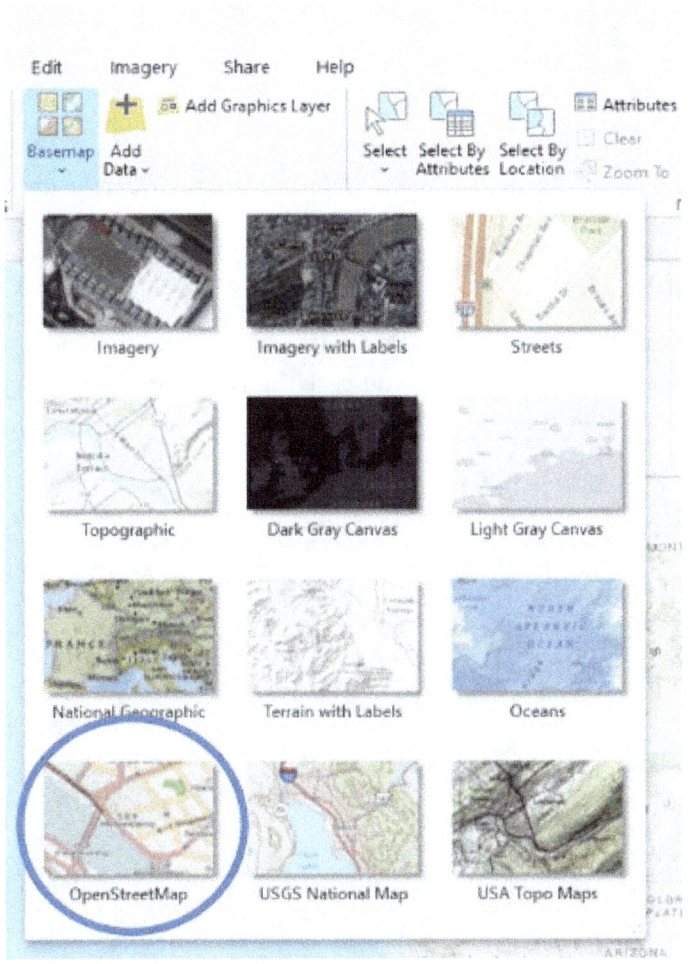

Section 2.1 Task: Submit a screenshot of your map on ArcGIS Pro with the OpenStreetMap basemap.

23

Section 2.2: Adding vector data

Vector data can be stored in three ways: points, lines, and polygons. Points are the most basic form of vector data. Many points connected together form a line, and many lines connected together form a polygon. Each data file has the features (points, lines, or polygons) with a related table that stores all the information about those features. This table is called an attribute table. There are different ways to analyze vector data for the different types of data that is available.

This part of the tutorial will demonstrate how to add the different types of vector data to your mapping project.

1. Continuing in the map with the OpenStreetMap basemap, you will now add the **CholeraDeaths** file to the interface. To do this, you click the **Add Data** icon along the top of the workspace under the **Layer** section.

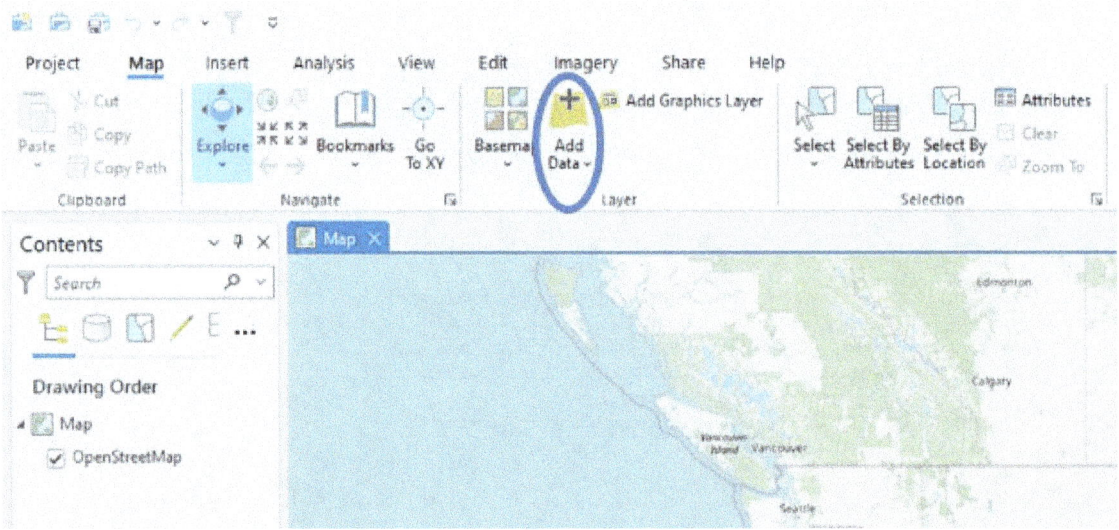

2. Next, you will need to navigate to the **JohnSnow.gdb** geodatabase that you downloaded with the tutorial data and select **CholeraDeaths** shapefile. Then click **OK**.

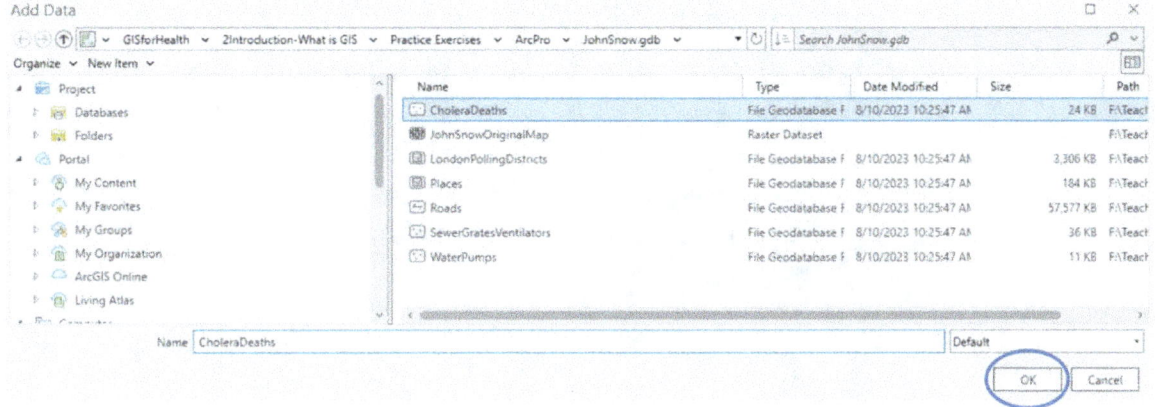

3. As you can see, this shapefile is a point location file. It stores the geographic locations of several points along with information about each point. You can see the information about each point when you open the **Attribute Table**. To do this, right click the layer name (**CholeraDeaths**) in the contents pane and click **Attribute Table**. Then review the data available for each of the points.

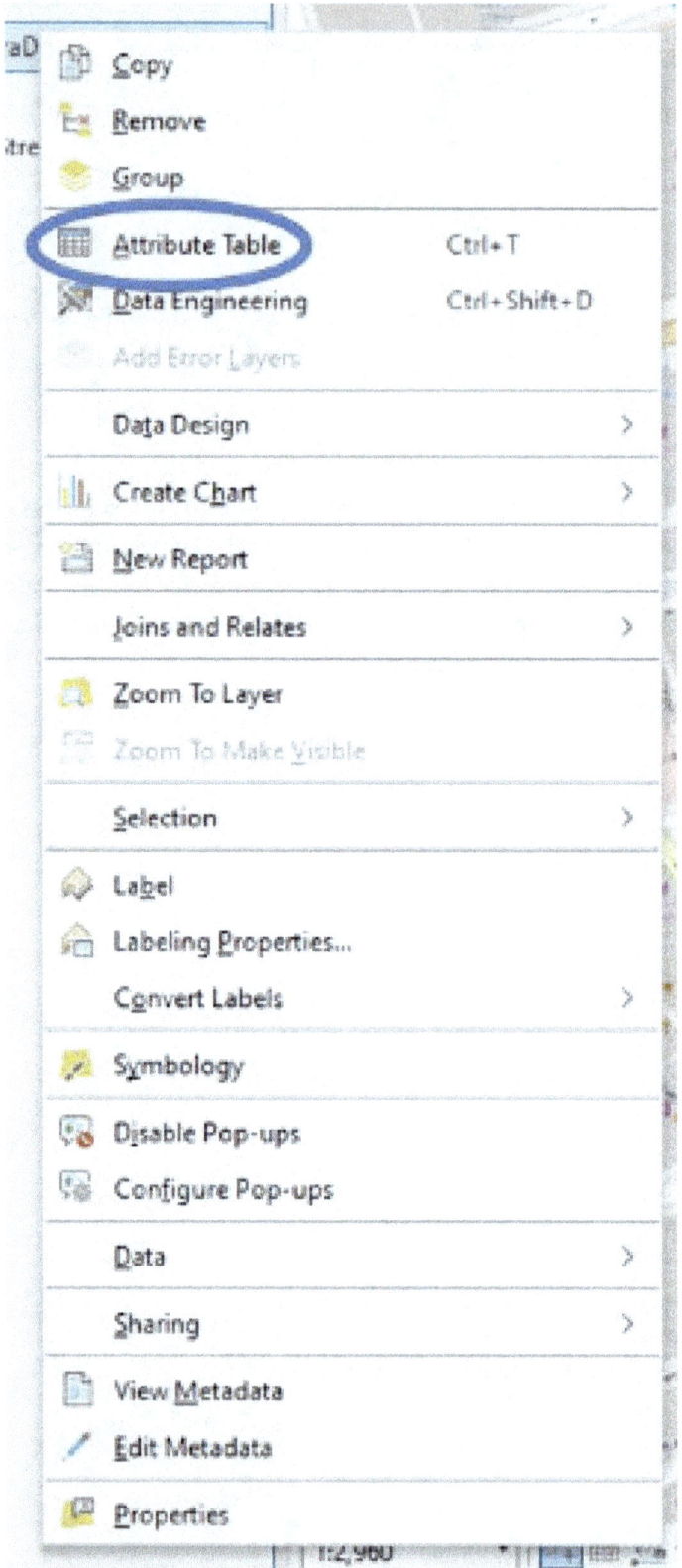

4. Next, we are going to add a line shapefile to your map called **Roads**. To do this, click the **Add Data** button again. You should still be in the **JohnSnow.gdb** folder. Here you will select the **Roads** shapefile and click **OK**. Open the **Attribute Table** and view the information that is provided for each of these lines.

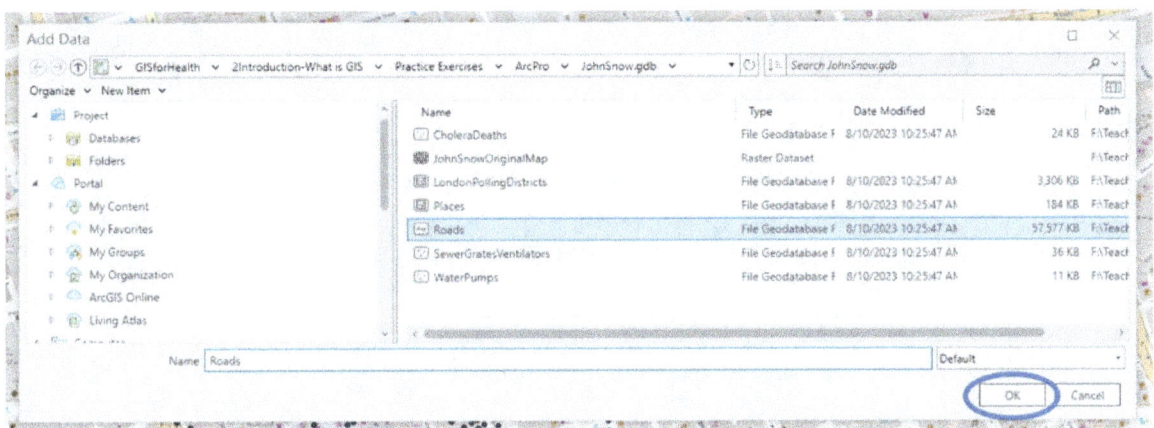

5. Finally, we will do the same process to import a polygon shapefile called **LondonPollingDistricts** into the map. Open the **Attribute Table** and view the information that is provided for each of these polygons.

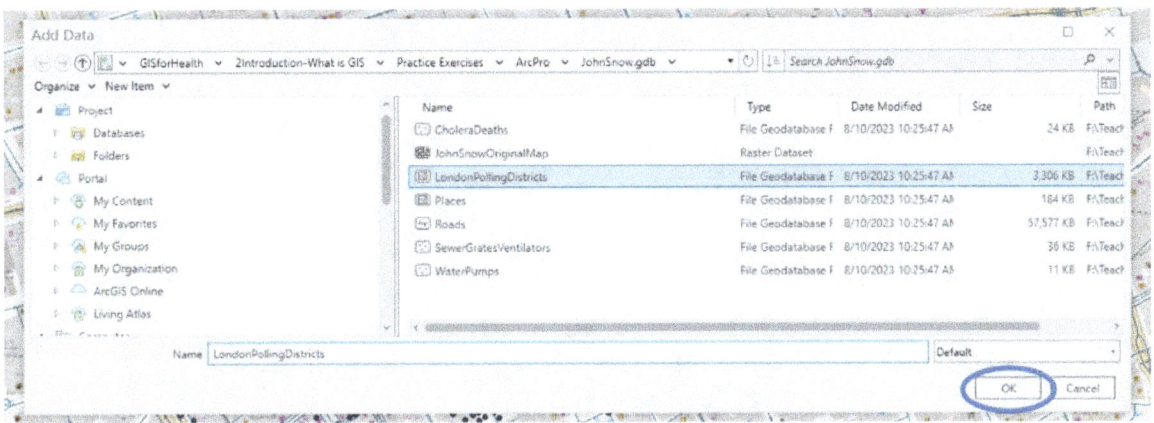

27

6. Next, we will learn how to change the extent you are viewing. You want to see the extent of all the data in your map. This can be done by zooming to your largest layer. To do this, you right click your largest layer (**LondonPollingDistricts**) and choose **Zoom To Layer**.

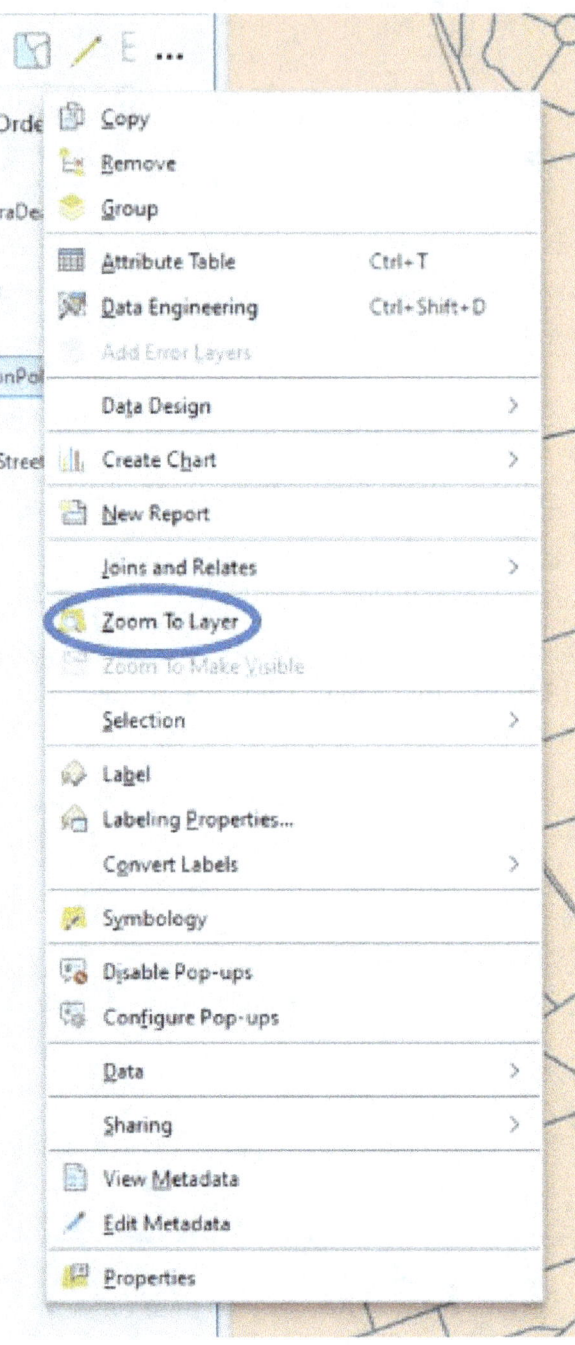

Section 2.2 Task: Submit a screenshot of your map on ArcGIS Pro of the full extent of your three vector layers added.

Section 2.3: Adding raster data

Raster data is made up of pixels. You can use any image that has a red, green and blue value. You can get images of the earth via satellites that can represent land cover or use old maps and georeference them. This process involves lining up the map to a basemap to give the image location data. You can also take vector data and transform it into raster data.

This part of the tutorial will demonstrate how to add an old map that has already been georeferenced into your mapping project.

1. Continuing in your map, you will now add John Snow's original Cholera outbreak map to the interface. To do this, you click the **Add Data** icon and select **JohnSnowOriginalMap**. Then click **OK**.

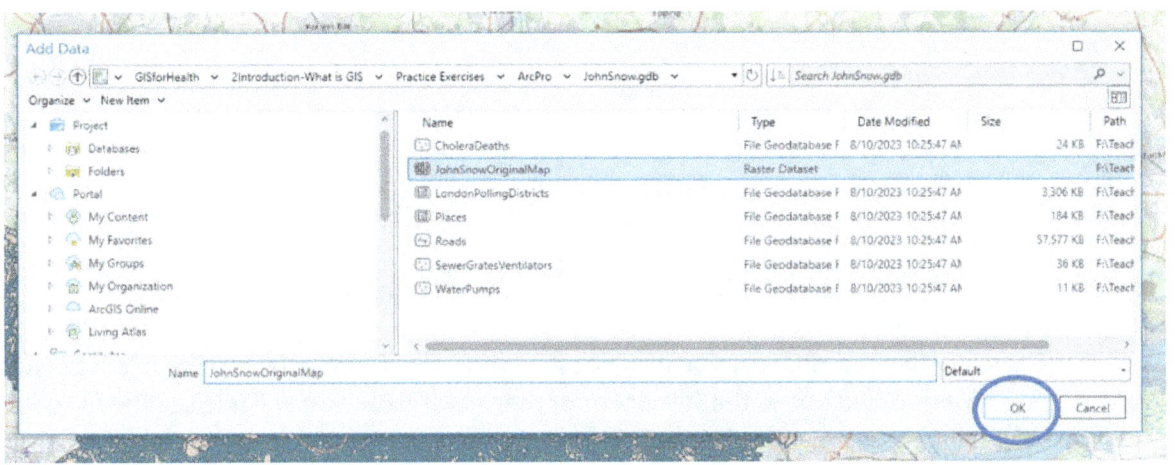

2. If you right click the layer, you will notice that the **Attribute Table** is grayed out. This is because raster data does not have ready attribute tables.

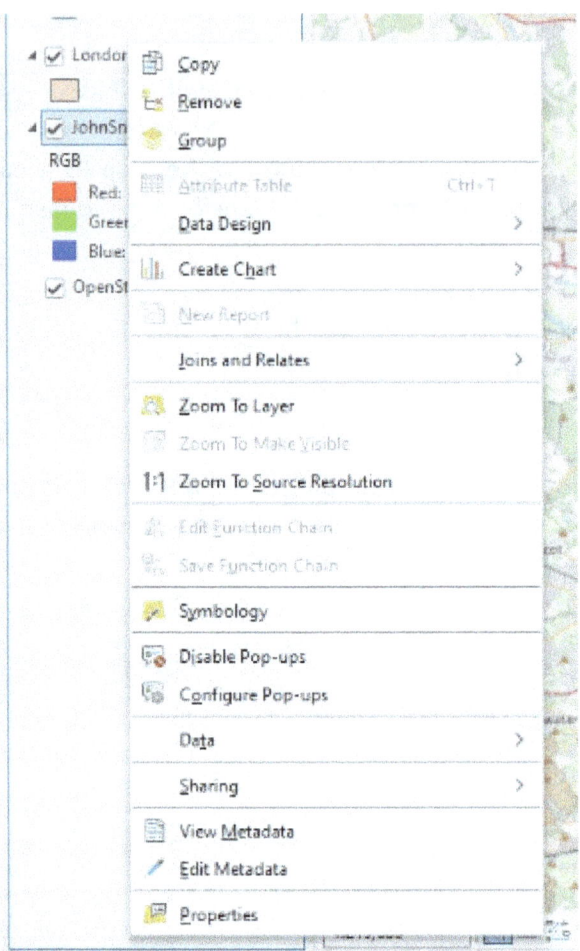

3. There are two reasons why you may not see John Snow's original map. The first is because you have zoomed out to the full extent of the vector data, you will not be able to see the map right away. You can zoom into the map by right clicking **JohnSnowOriginalMap** in the contents pane and selecting **Zoom To Layer**. If you still cannot see the map, you will have to rearrange the layers in the content pane. To do this, click and hold **JohnSnowOriginalMap** and drag it to the top of the list.

Section 2.3 Task: Submit a screenshot of your map on ArcGIS Pro of the zoomed into **JohnSnowOriginalMap** with the original map at the top of the drawing order.

Section 3: Working in the ArcGIS Environment
Section 3.1: View and edit metadata for layers in ArcGIS Pro

Metadata, put simply, is data about data. This stores information such as what the data is, when it was developed, and who created it. It is very important to keep up with metadata for spatial data, especially if the data is to be shared in the future with other people or groups.

This part of the tutorial will demonstrate how to view and edit metadata for layers in ArcGIS Pro.

1. Continuing in your map, you will now look at the metadata for each of the layers. To do this, you right click the layer in question and select **View Metadata**. Start with **CholeraDeaths**.

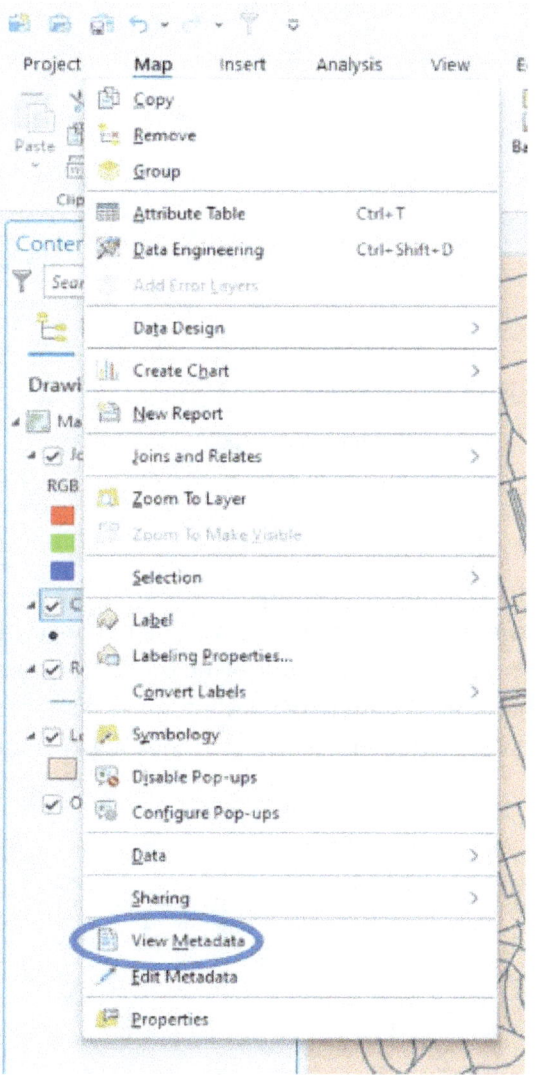

You will notice that you can switch between the layers in the **Catalog** tab that has popped up. While looking at **CholeraDeaths**, you can see that it has a Title along with information about its Type, Tags, Summary, Description, Credits, Use limitations, Extent, and Scale Range. The metadata for this shapefile has been updated and maintained as recommended. Now begin looking at the metadata for the other layers.

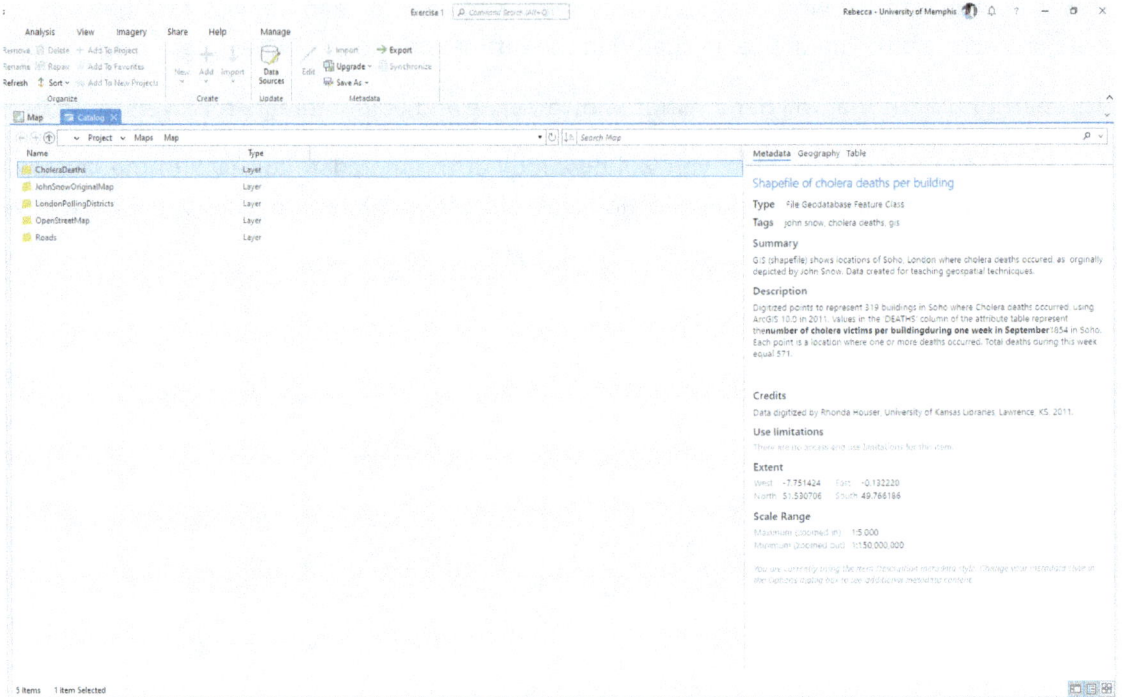

1. You will notice that some of the other layers are not as well kept. We must enter in Metadata for these files. To do this, we must first change the layer's properties to assign the layer its own metadata instead of copying the source metadata. First, close the **Catalog** tab. Then, right click **JohnSnowOriginalMap** in the contents pane and select **Properties**. Then, from the drop down menu at the top, choose **Layer has its own Metadata** and click **OK**.

32

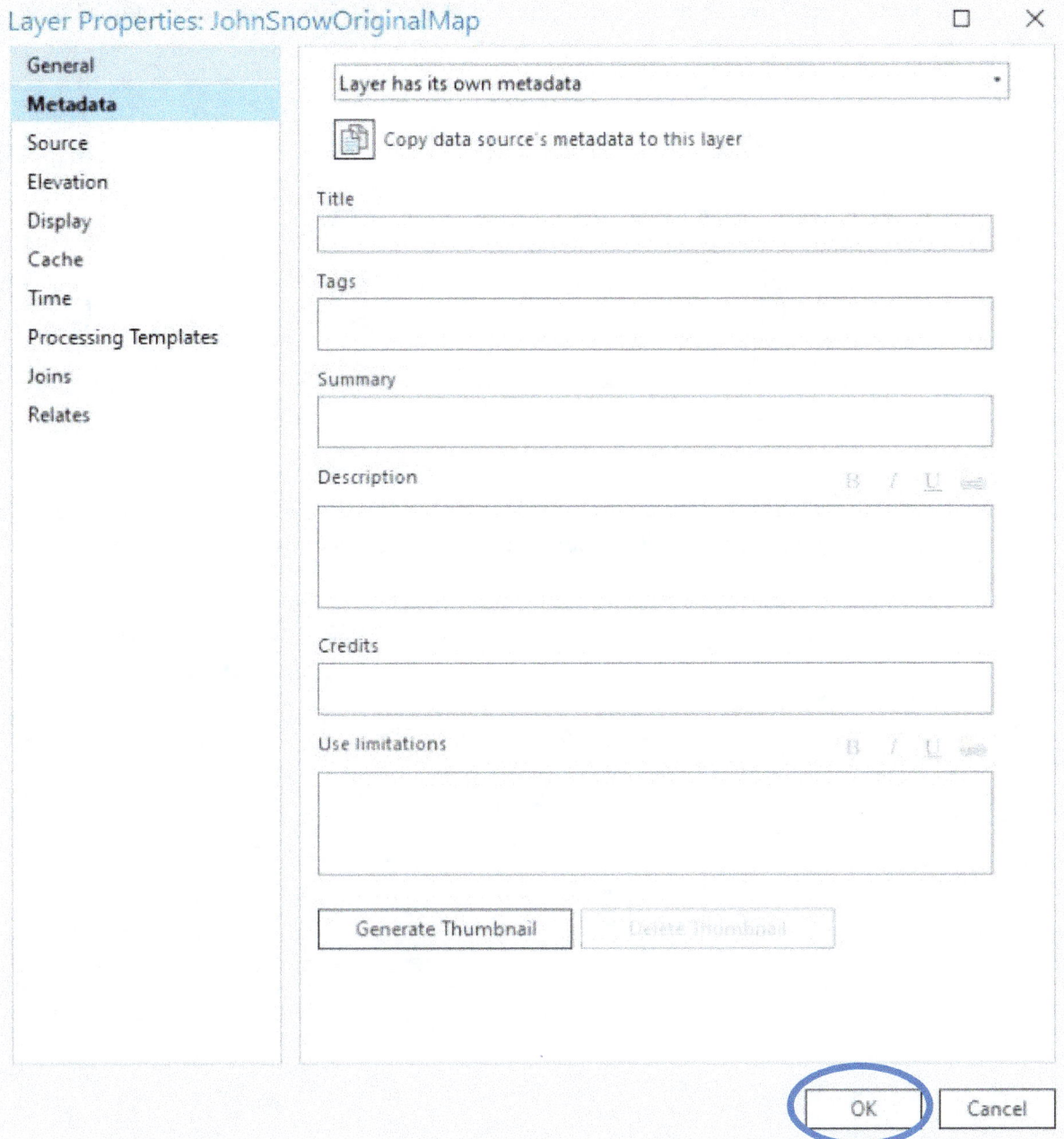

2. Next, click the **View** tab at the top of the ArcGIS Pro interface. Then click **Catalog View**.

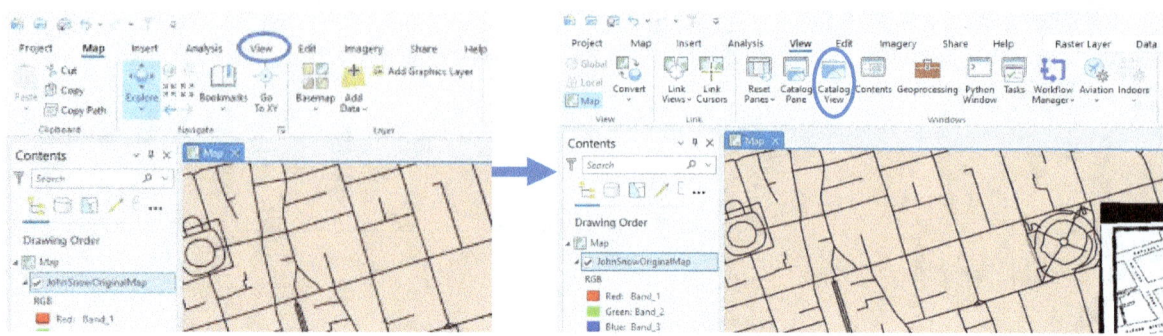

3. Next, you will need to navigate to the folder for the **JohnSnow.gdb**. You can then edit the metadata by right clicking **JohnSnowOriginalMap** and selecting **Edit Metadata**.

4. Fill in the metadata according to the figure below then close the **JohnSnowOriginalMap** tab. It should ask you if you want to save your changes. Click **Yes**.

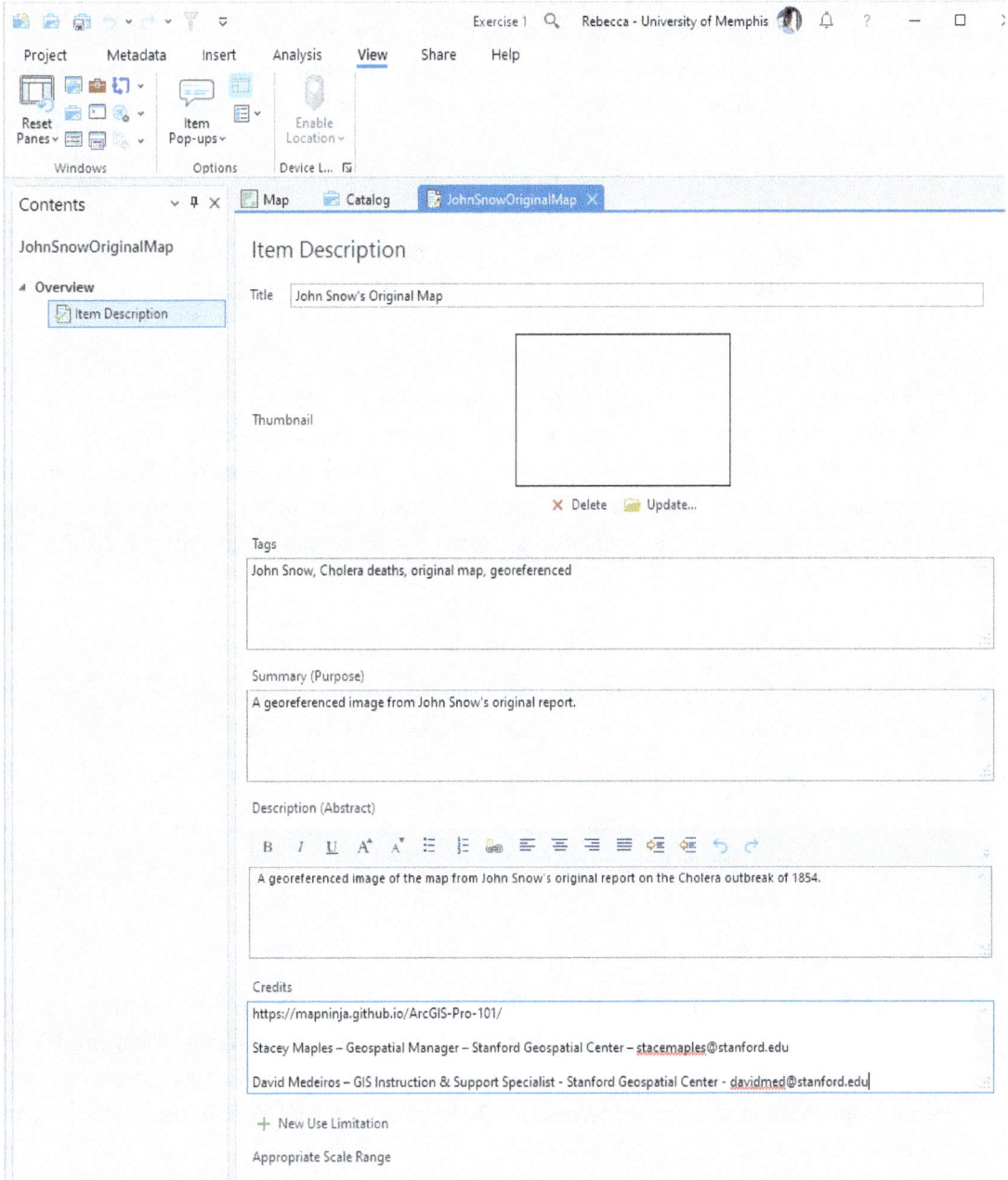

5. In Catalog **View**, you will now see the newly entered metadata if you click on the **JohnSnowOriginalMap** layer.

Section 3.1 Task: Submit a screenshot of the metadata you entered for the **JohnSnowOriginalMap**.

Section 3.2: Select certain features from a layer

One thing that will make the finished maps look crisp and clean is making sure that all the layers you have added to the map cover the same area. There are a few tools that can help you select certain parts of a layer (the **Select** tools). Then you can create temporary layers from those selected features.

This part of the tutorial will demonstrate how to select specific features from a layer using the **Select** tools in ArcGIS Pro. Then create a temporary layer of those selected features.

1. Continuing in your map, you see that there are two layers that reach much further throughout the map than the rest: **Roads** and **LondonPollingDistricts**. We want to reduce these two layers to show only our areas of interest. There are three helpful tools that allow us to select features from a layer: **Select**, **Select By Attributes**, and **Select By Location**. They are all located in the **Selection** section of the toolbar at the top of the ArcGIS Pro environment.

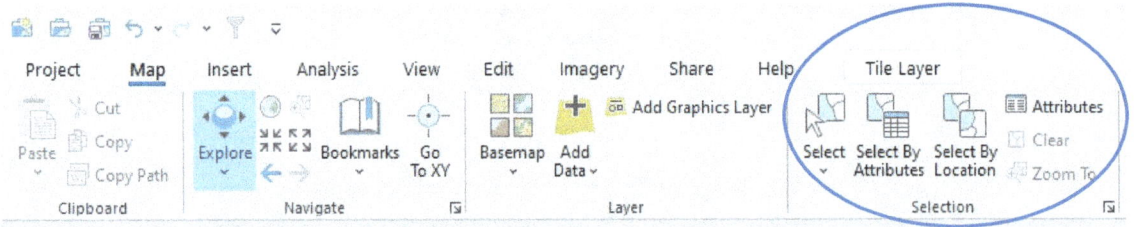

2. Let's start by narrowing down the polling districts to only those found within the area of interest. First, click the **Select By Attributes** tool. Then, in the **Input Rows** section select **LondonPollingDistricts**. Using the drop-down menu next to **Where**, select **PD_ID** and search for **WEA** in the drop down menu next after **is equal**. It will look according to the figure below.

36

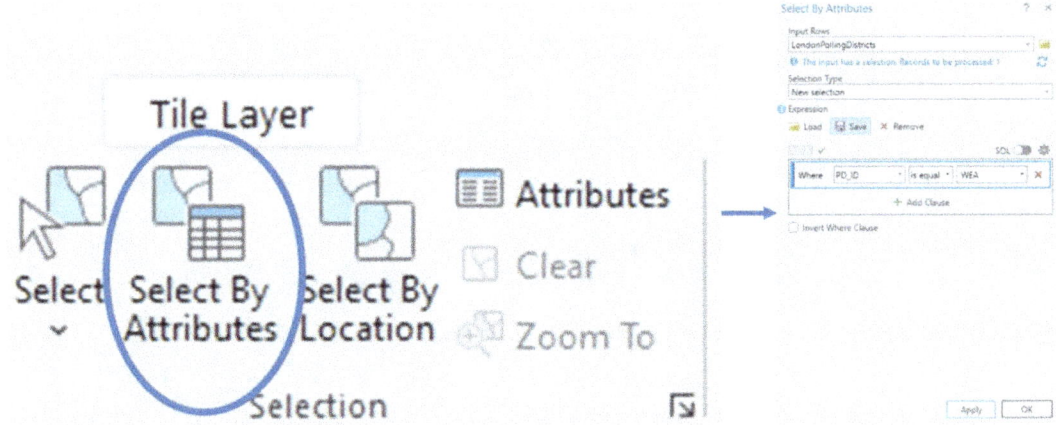

3. Next, click the **Apply** button to apply the selection. This will select only the features whose **PD_ID is equal to WEA**. These features will have a colored ring around them to show that they are selected. It may be hard to see which features are selected. You can zoom into those selected features by right clicking the layer (**LondonPollingDistricts**), hover over **Selection**, then click **Zoom to Selection**. NOTE** In the figure below, it is highlighted in red, yours may be highlighted in a different color such as blue.

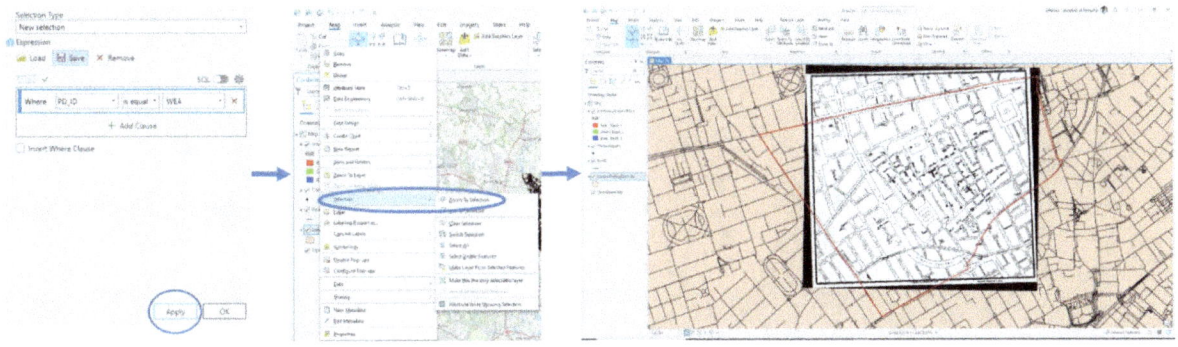

4. Next, you will want to make a temporary layer out of the selected features. To do this, by right clicking **LondonPollingDistricts**, hovering over **Selection**, and clicking **Make Layer From Selected Features**. The temporary layer automatically gets added to the top of the drawing order in your map and it will be named **LondonPollingDistricts selection**.

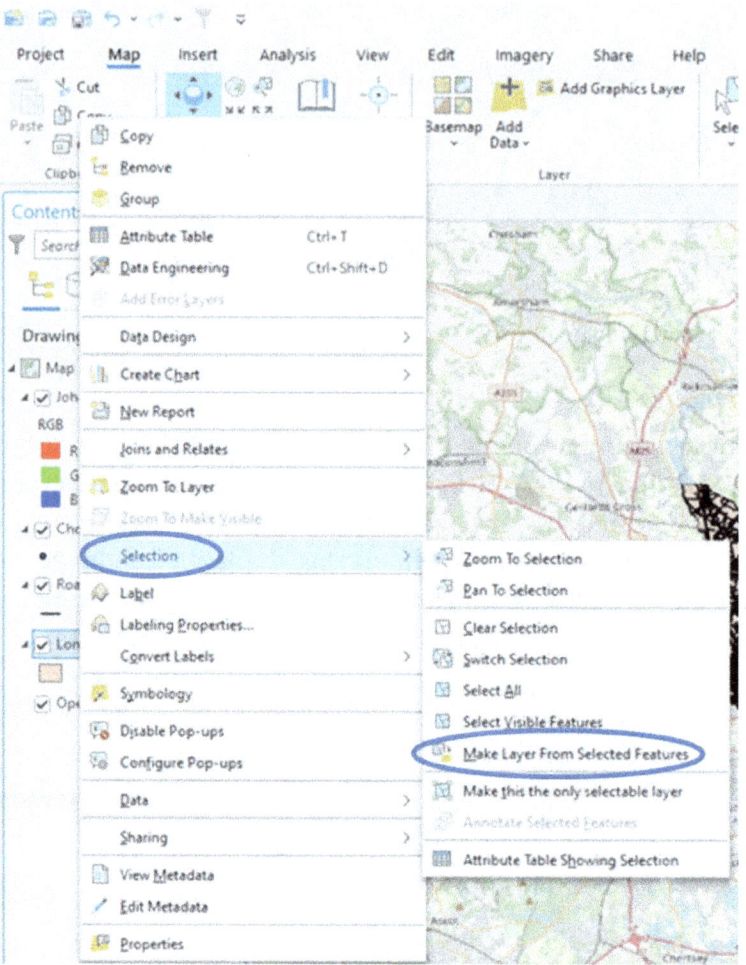

5. Now it is time to select only the roads in the area of interest. Before working on this, you must clear the selected features from before. Choose the **Clear** tool in the **Selection** section at the top of the ArcGIS Pro work environment.

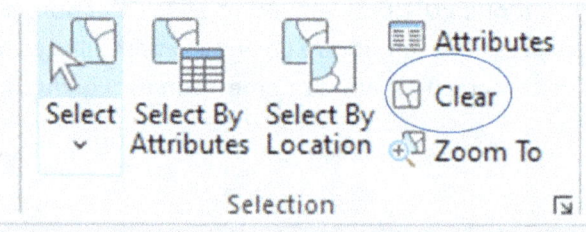

6. To do this, choose the **Select By Location** tool in the **Selection** section at the top of the ArcGIS Pro work environment. Then, in the **Input Features** section make sure **Roads** is selected. Under the **Relationship** section, use the drop-down menu to select **Within**. Next, use the drop-down menu under **Selecting Features** to choose **LondonPollingDistricts selection**. When finished it will look according to the figure below.

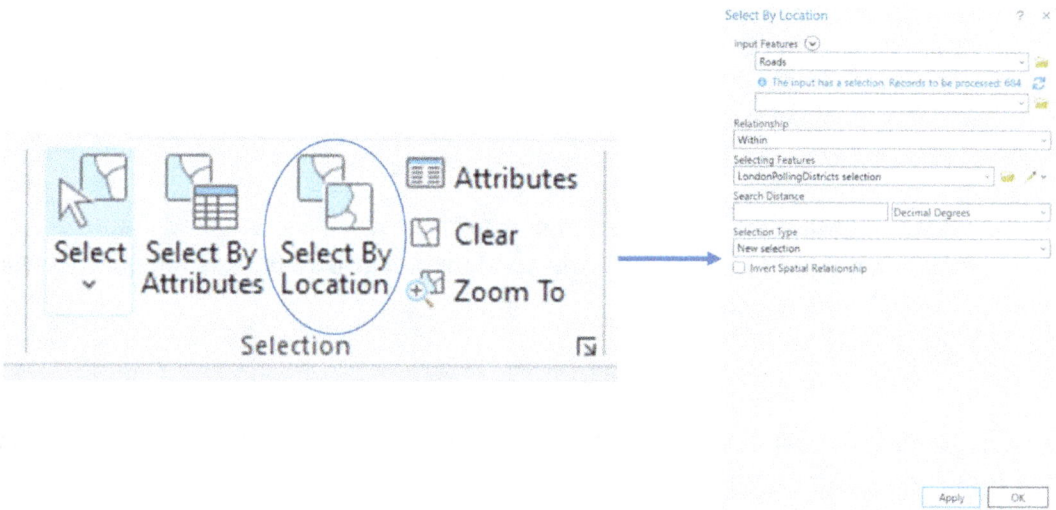

7. Next, click the **Apply** button to apply the selection. Then click **OK**. This will select only the features in the **Roads** layer that fall **within** the outline of the **LondonPollingDistricts selection** layer. These features will be highlighted to show that they are selected. **NOTE**** In the figure below, it is highlighted in red, yours may be highlighted in a different color such as blue.

39

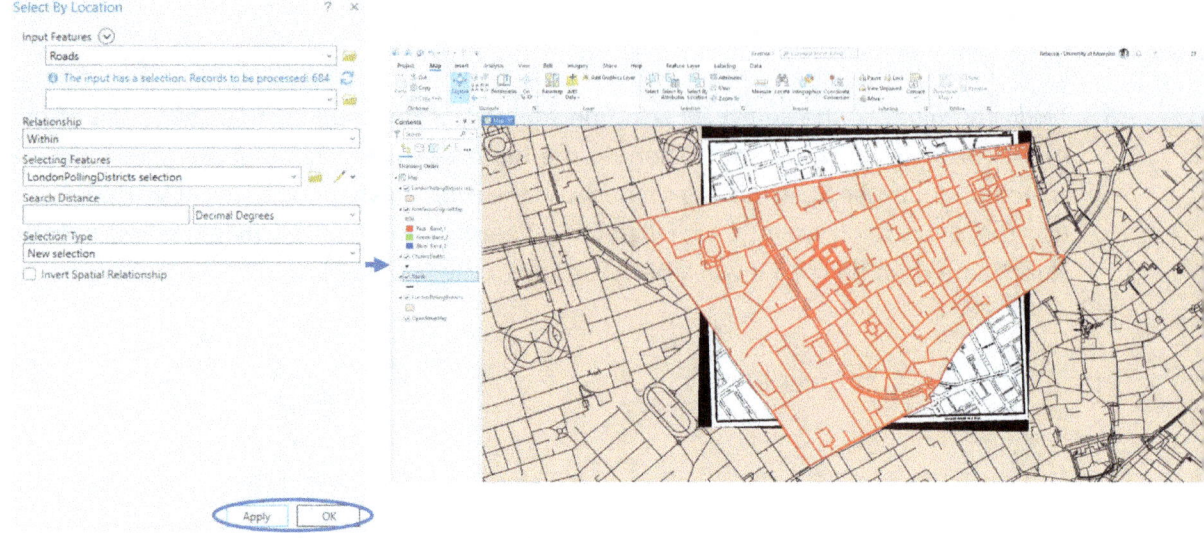

8. Next, you will want to make a temporary layer out of the selected features in the same manner as when working with the **LondonPollingDistricts**. This temporary layer will be named **Roads selection**.

Section 3.2 Task: Submit a screenshot of your map with the **LondonPollingDistricts selection** and **Roads selection** layers at the top of the drawing order.

Section 3.3: Export spatial data and maps

One of the main areas of science that generally gets overlooked is sharing your work with others. In geospatial sciences, you can share your work with others by exporting the spatial data you have worked on as new shapefiles or by exporting the finished maps to allow people a visual representation of the data.

This part of the tutorial will demonstrate how to export spatial data and maps in ArcGIS Pro.

1. You had created a layer from selected features that you may want to use in the future in another map, or others may want to use later. To be able to use it in future maps, you must export the temporary layer to make a permanent shapefile in your geodatabase. You can export the data to create a permanent layer within your geodatabase. To do this, you

will right click the temporary layer (**LondonPollingDistricts selection**), hover over **Data** and click **Export Features**.

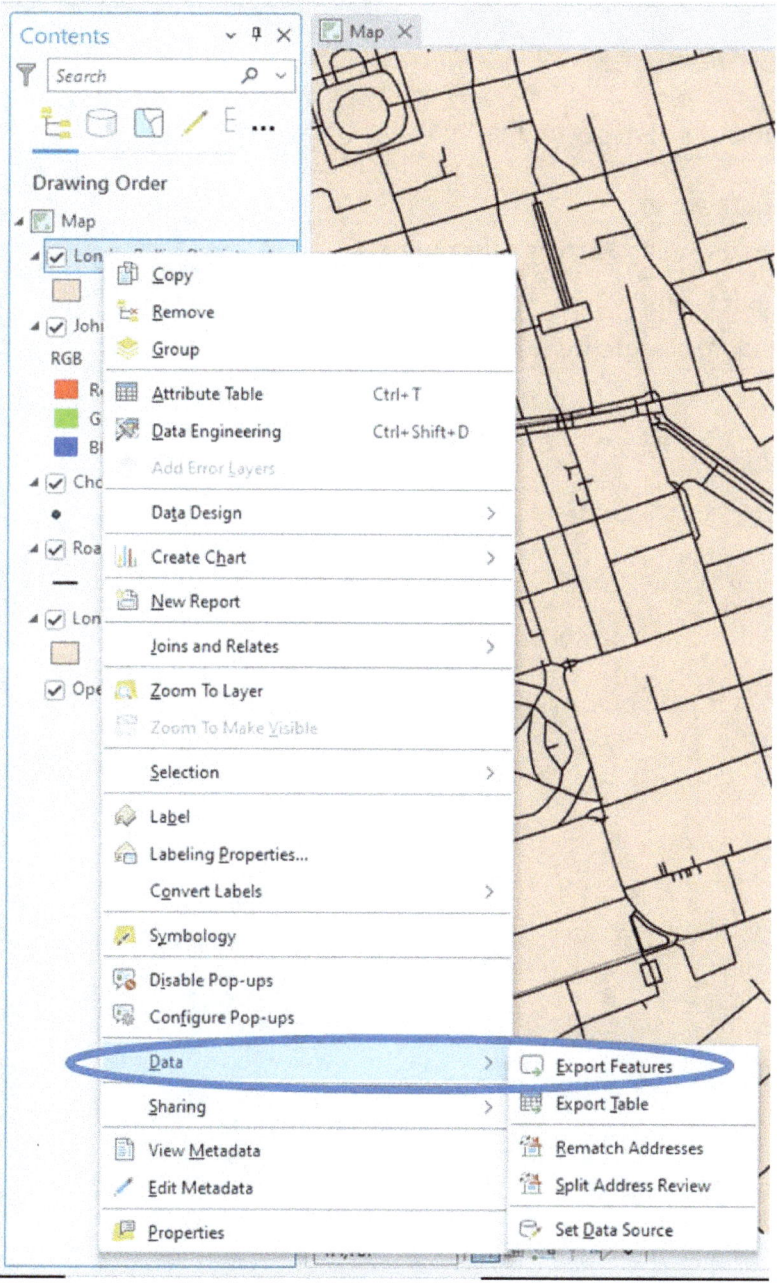

2. A window will pop up with information to fill out. Here you keep the **Input Features** the same. For the **Output Feature Class**, click the folder icon towards the right side of the entry box.

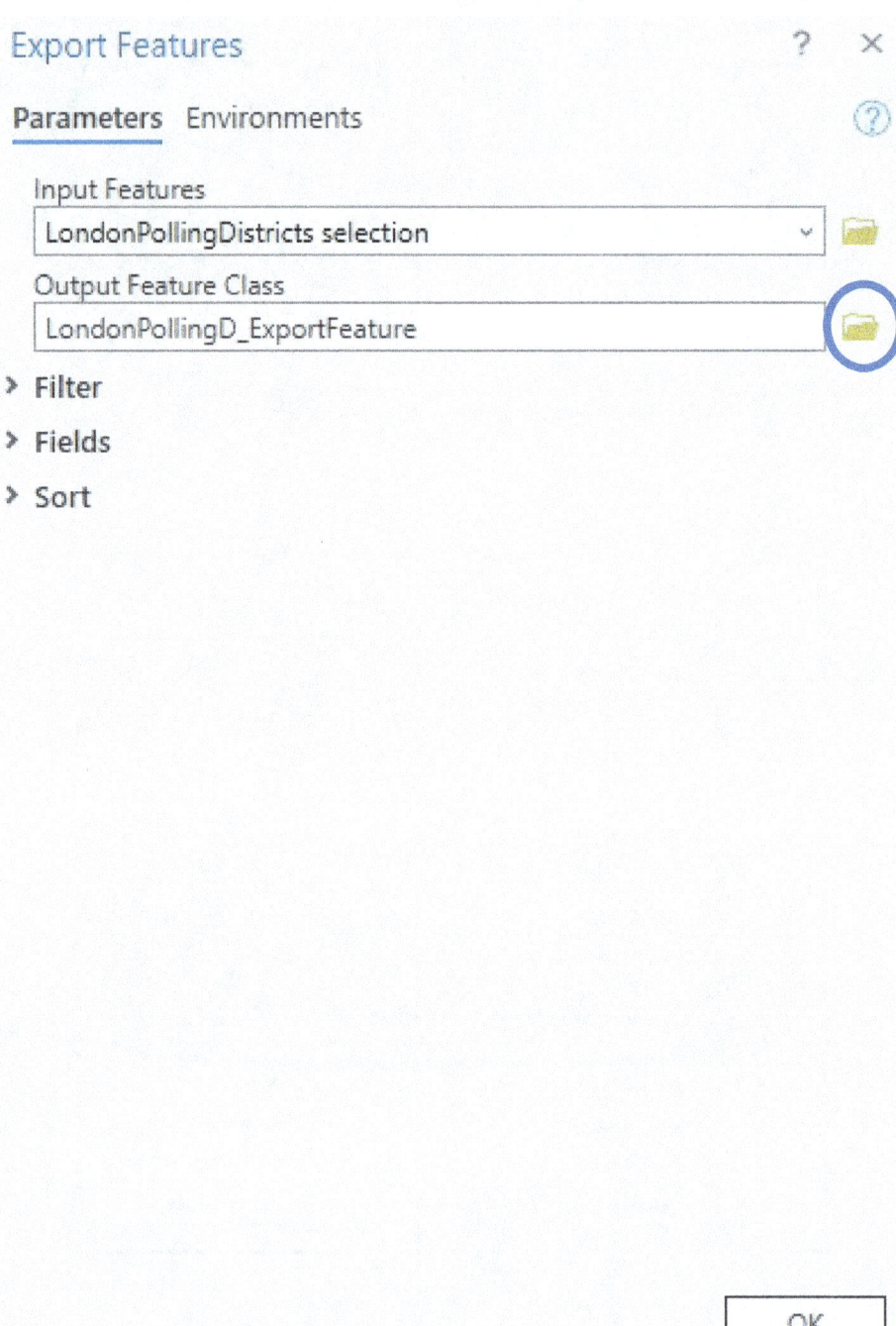

3. Navigate to the folder for the **JohnSnow.gdb** and name the layer **StudyArea**. Click **Save**, then **OK**. You now have a new permanent shapefile that outlines your study area in your geodatabase. You will notice that the new permanent shapefile automatically gets added towards the top of your map.

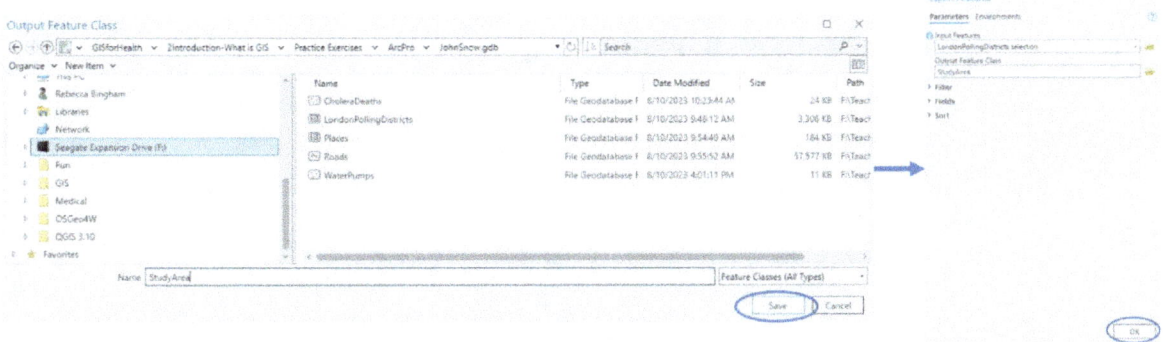

4. Do this for the **Roads selection** layer as well using the information below in the **Export Features** window.

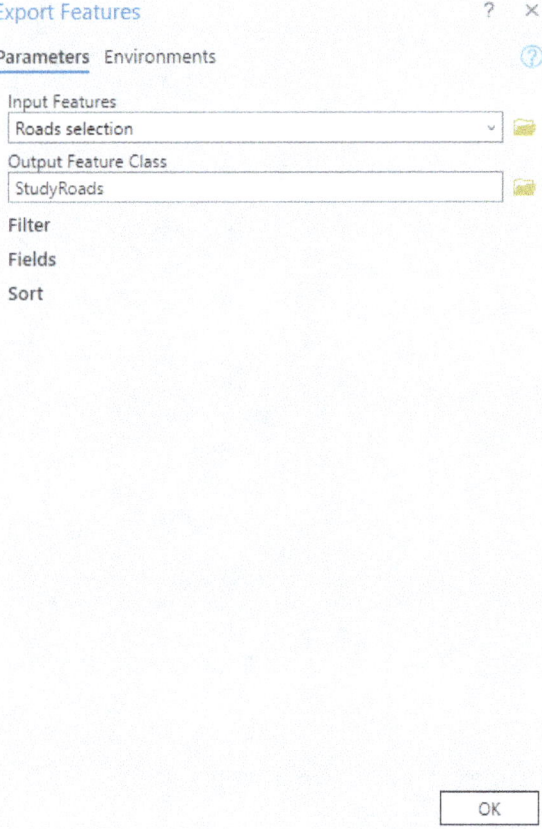

5. Move **CholeraDeaths** to the top of the drawing order. Uncheck the boxes next to the other layers besides **CholeraDeaths**, **Roads**, and **StudyArea** to hide them.

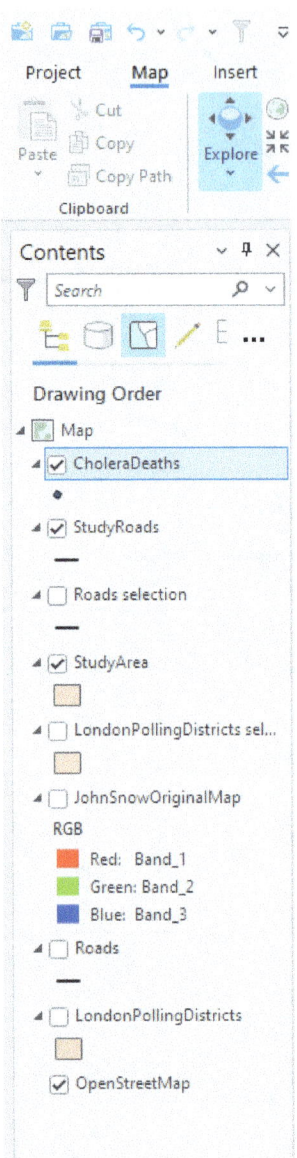

Section 3.3 Task: Submit a screenshot your map.

CHAPTER 2: SPATIAL ANALYSIS AND HEALTH

Mastering the Concepts

This chapter explores spatial analysis, a powerful method for examining geographic patterns and relationships through spatial data. It highlights how this approach is particularly valuable in understanding the influence of the environment on various phenomena, with a focus on public health. By utilizing tools such as maps, geographic information systems (GIS), and statistical techniques, spatial analysis enables the exploration of spatial patterns, identification of trends, and interpretation of complex data. For health professionals, integrating spatial data helps connect specific locations to health outcomes, revealing insights that traditional analysis alone may overlook. This approach is crucial for public health, as it provides a deeper understanding of how environmental and social factors intersect with human health.

One of the most famous examples of spatial analysis in public health history is John Snow's investigation into the 1854 cholera outbreak in London. Snow is considered one of the pioneers of spatial analysis in public health. During the outbreak, he mapped the locations of cholera deaths in relation to public water pumps. Snow discovered that the majority of deaths were clustered around a particular pump on Broad Street. This observation led him to hypothesize that cholera was transmitted through contaminated water, challenging the prevailing theory that the disease was spread by "bad air" or miasma. Snow's use of spatial analysis to identify the source of the cholera outbreak is often credited as the birth of modern epidemiology and the use of maps to track and understand disease patterns.

Geographic Information Systems (GIS) play a central role in spatial analysis by providing a framework for storing, managing, analyzing, and visualizing spatial data. GIS tools enable health professionals to map and spatially integrate different types of public health data, such as disease incidence, environmental hazards, and healthcare access. By using GIS, health professionals can identify spatial patterns and correlations that might otherwise go unnoticed, helping to uncover links between health outcomes and environmental factors. This allows for more targeted interventions and resource allocation, leading to improved health outcomes for communities. GIS allows spatial data to be processed, visualized, and analyzed to make informed decisions and recommendations.

For health professionals, spatial analysis is a powerful tool that enables them to understand the complex relationships between geography, environmental conditions, and public health. By applying spatial analysis techniques, health professionals can gain a deeper understanding of the environmental and social determinants of health and use this information to make data-driven decisions. Social determinants of health, such as income, education, employment, housing, and access to healthcare, are influenced by geographic factors. Through spatial analysis, we can better assess how these factors cluster in certain areas and how they correlate with health disparities.

For instance, spatial analysis can help public health professionals assess access to healthcare services, identify areas of health inequity, track the spread of infectious diseases, and analyze environmental exposures like pollution or pesticides. By mapping access to medical care and food deserts, for example, spatial analysis reveals where resources are scarce, and communities face greater challenges. This spatial lens allows for targeted interventions and more equitable resource allocation, addressing the root causes of health disparities.

Spatial analysis can be used to characterize the environment by understanding how various environmental conditions—such as climate, water quality, air pollution, poverty, and traffic patterns—affect public health. Social determinants like poverty and housing conditions can be mapped alongside disease prevalence, revealing correlations between socioeconomic status and health outcomes. For example, health professionals can map areas with poor air quality and higher rates of respiratory diseases, providing evidence to inform air quality regulations or policies. Similarly, spatial analysis helps assess access to resources like healthcare, nutritious food, and safe living environments. By mapping the proximity of communities to healthcare facilities or food deserts, health professionals can identify areas where interventions or infrastructure improvements are most needed.

Another important application of spatial analysis is identifying health risks. By mapping the spread of infectious diseases, such as malaria, Lyme disease, and tuberculosis, spatial analysis can reveal geographic areas that are at higher risk of outbreaks. This can help health professionals anticipate future cases and deploy resources effectively. Additionally, spatial analysis can identify potential cancer clusters or areas with higher rates of environmental exposures that may contribute to health risks. For instance, health professionals might map areas with high concentrations of industrial pollution and correlate them with incidences of certain cancers, helping to identify environmental factors that could be impacting public health.

Spatial analysis also plays a critical role in linking health outcomes to exposures. Through the use of GIS, health outcome data (such as disease rates, mortality, or life expectancy) can be linked to environmental exposure data (such as pollution levels, proximity to hazardous waste sites, or traffic patterns). This enables health professionals to identify spatial correlations between exposures and health outcomes, helping to uncover environmental risks that may be contributing to poor health in certain populations. Moreover, GIS tools can be used to examine how socio-economic factors like income or education influence health outcomes, providing deeper insights into the social determinants of health.

Spatial analysis continues to play an important role in public health today, helping health professionals identify trends, assess risks, and target interventions more effectively. One notable example of spatial analysis in public health today is the use of GIS to monitor and control infectious diseases. For example, during the Ebola outbreak in West Africa, GIS was used to map the spread of the virus, identify high-risk areas, and plan intervention strategies. Similarly, GIS has been used to track the spread of COVID-19, allowing public health authorities to map the distribution of cases and implement containment measures based on geographic patterns.

More and more universities are creating certificate programs that combine GIS, spatial analysis, and public health. These programs teach students how to use spatial analysis tools to better understand the relationship between geography and public health. Students in these programs learn to synthesize geographic data with public health information, allowing them to identify how factors such as urbanization, migration, environmental hazards, and public health infrastructure affect health outcomes. Students also gain the skills to collect, analyze, and visualize spatial data using GIS, which enables them to interpret health outcomes and design interventions based on geographic patterns.

Through the application of spatial analysis, health professionals can improve their understanding of the complex relationships between geography, environmental conditions, and health. This approach provides valuable insights into how social and environmental factors contribute to health outcomes and allows health professionals to develop more effective strategies for disease prevention, resource allocation, and improving public health. As demonstrated by John Snow's work, spatial analysis is an essential tool for uncovering patterns, informing public health interventions, and ultimately improving the health and well-being of communities.

Mastering the Skills

Exercise 2: Visualizations in ArcGIS Pro – Learning basic tools and how to visualize spatial data

In this tutorial you are going to become familiar with some basic tools used in the ArcGIS environment to help organize and manipulate your data allowing for initial spatial data exploration. Then you will learn how to visualize the spatial data by understanding how to use colors, change the graphic elements on the map, and exploring data visualization over different categories and numeric scales. We are going to be working with data from Dr. John Snow's famous Cholera outbreak map. Read some background information about this event here: https://www.ph.ucla.edu/epi/snow/snowcricketarticle.html.

OBJECTIVES

- Open an ArcGIS Pro project file
- Set Data Source
- Spatial join
- Creating a buffer
- Changing symbol graphic elements
- Graduated symbols
- Choropleth maps
- Creating a heat map

Required data:

The original data sources listed below are provided for reference only. You do not need to download or curate these datasets from the original sources. All data, preconfigured GIS project files, and geodatabases required for this exercise are already included with the tutorial data and are available through the Resources page of the author's official website:

www.esraozdenerl/resources. Instructions for accessing the tutorial data can be found in the Tutorial Data section of the Preface.

Original data sources:

2. Data included in the map project file (same as exercise 1). Sources:
    a. Cholera deaths and water pumps (Point data):
       https://kuscholarworks.ku.edu/handle/1808/10772
    b. Road and place boundary shapefiles for the Greater London Area (Line and polygon data): http://download.geofabrik.de/europe/great-britain/england/greater-london.html
    c. Polling districts for the Greater London Area (Polygon data):
       https://osdatahub.os.uk/downloads/open/BoundaryLine
    d. John Snow's Original Map Georeferenced (Raster data):
       https://github.com/mapninja/ArcGIS-Pro-101/tree/master

Section 1: Learning Basic Tools

Section 1.1: Open an ArcGIS Pro project

In the last exercise, you learned how to save an ArcGIS Pro project. This allows you to be able to readily open a map you have previously worked on and view, make edits, or share it.

For the first part of this tutorial, you will learn how to open the ArcGIS Pro project.

1. Open ArcGIS Pro on your desktop and make sure that you are signed into your ArcGIS Pro profile. If the project is one that you have recently worked on, you will see the name of it under **Recent Projects**. If not, you can click **Open another project**.

2. After clicking **Open another project**, a pop-up window will appear. Navigate to your tutorial data folder and select **JohnSnow.aprx** and click **OK**.

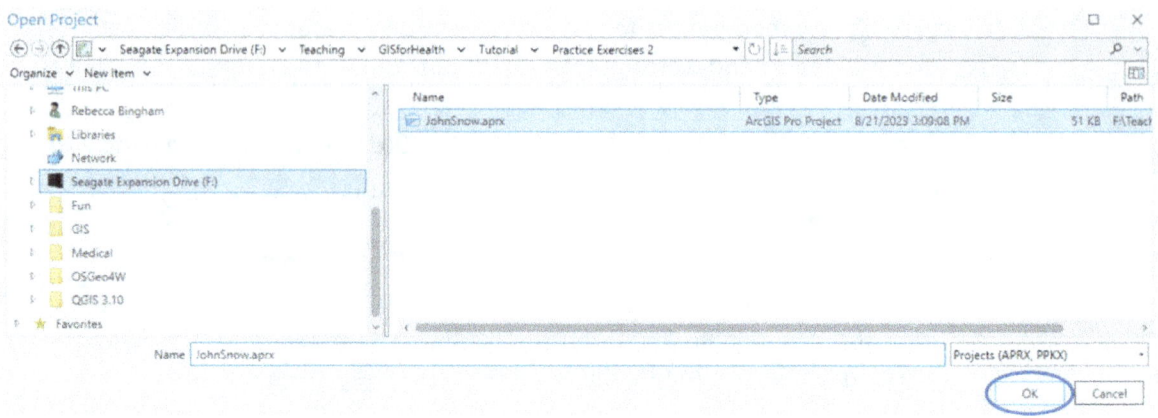

3. Once you have finished this, the software will open the **JohnSnow.aprx** mapping project, which contains four GIS vector layers over an OpenStreetMap basemap.

Section 1.1 Task: Submit a screenshot of the open map project on ArcGIS Pro.

Section 1.2: Set Data Source

Whenever you receive a map project from someone else, whether the data is included or you already have it, you should set the data source before you start working with the map. If the data path is broken, your project file may show data layers with red exclamation marks, indicating that you need to set the dataset. Setting the data source tells the map where your data is located and is essential for the map to function properly and display all layers correctly.

For this part of this tutorial, you will learn how to set the data source for a layer.

1. Right click the **WaterPumps** layer and choose **Properties**.

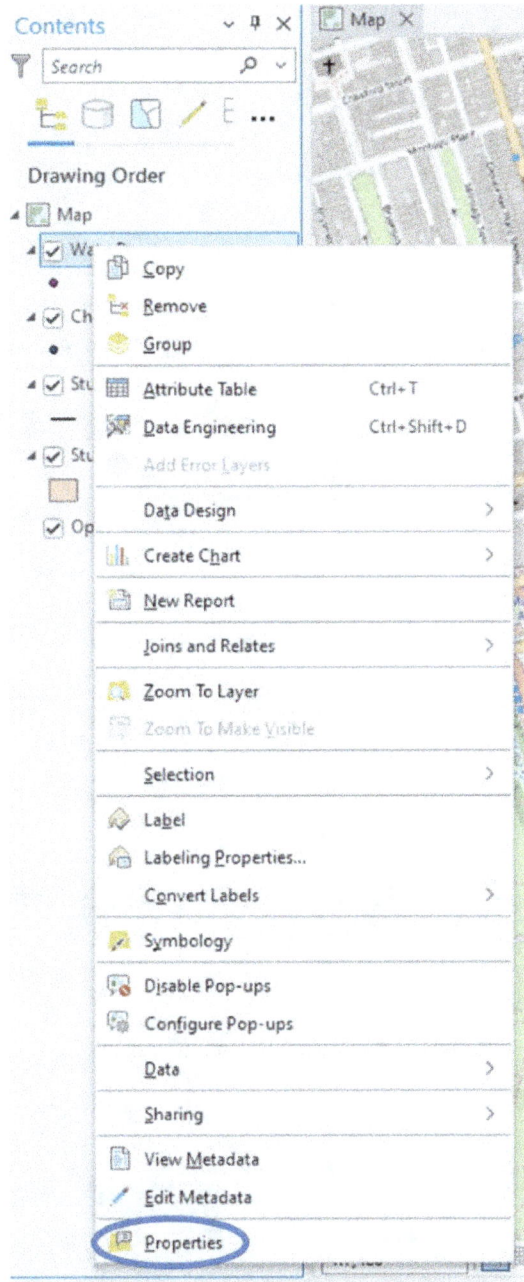

2. After clicking **Properties**, a pop-up window will appear. Select **Source** from the left-hand side. Then click **Set Data Source** on the top right-hand side.

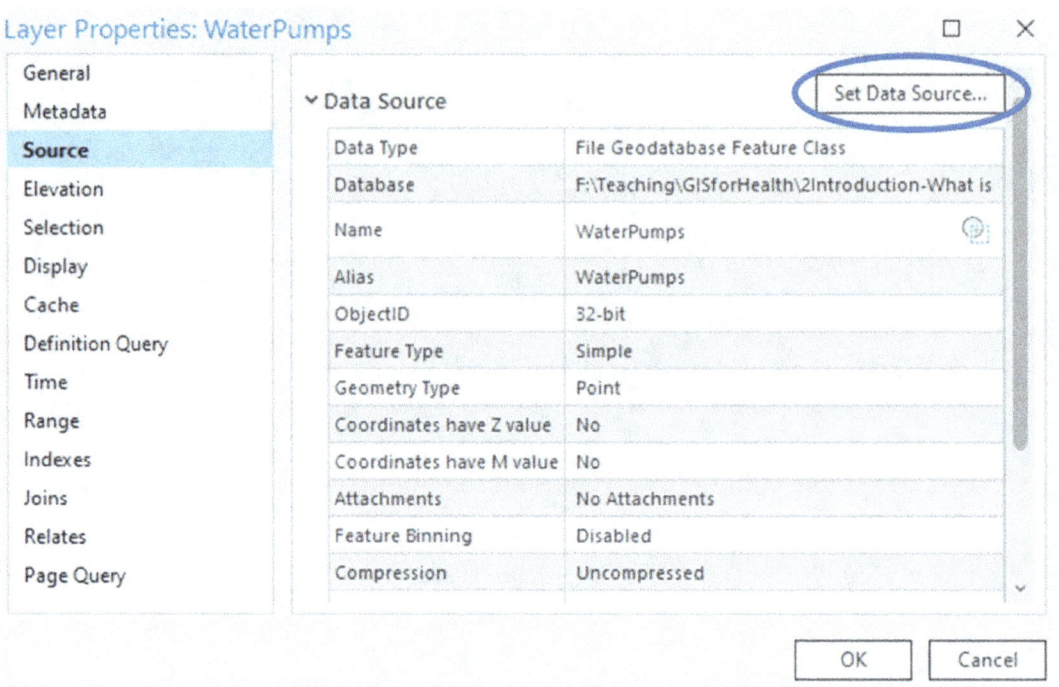

3. After clicking **Set Data Source**, another pop-up window will appear. Navigate to the **JohnSnow.gdb** geodatabase that you used in exercise 1. Then choose **WaterPumps**. Next click **OK.**

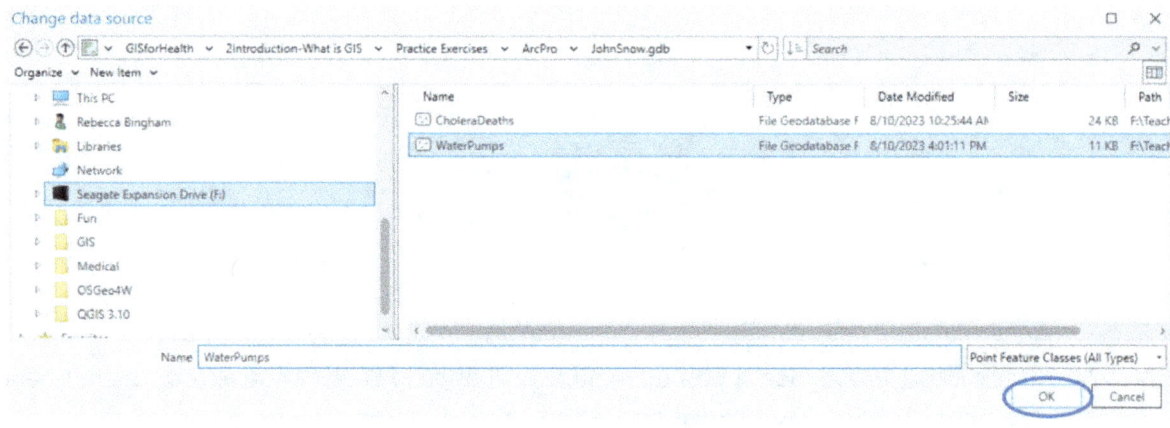

4. Follow the same steps for each of the other three data layers: **CholeraDeaths**, **StudyRoads**, and **StudyArea**. This will allow you to work with and view the data for this map project.

Section 1.2 Task: Submit a screenshot of your map with all data layers having their data source set.

Section 1.3: Spatial join

In some cases, you want to understand how one GIS layer relates to another spatially. For instance, if you want to understand what county a specific building is in or what road is closest to a certain point. You can assign these features to that building or road using a spatial join. The **Spatial Join** tool will match the rows from a join feature to a target feature based on their relative spatial locations.

For this part of this tutorial, you will learn how to use the spatial join tool.

1. Right click the **CholeraDeaths** layer and hover over **Joins and Relates**. Select **Add Spatial Join**.

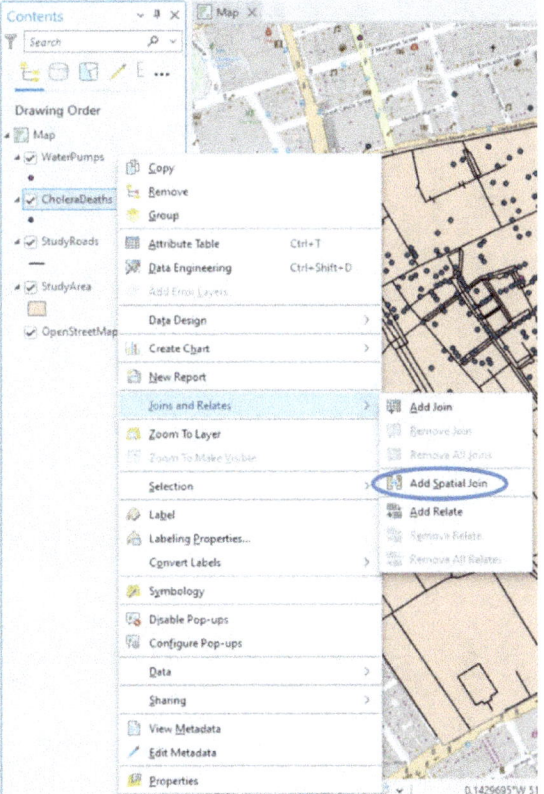

2. After clicking **Add Spatial Join**, a pop-up window will appear. Click the drop-down arrow under **Join Features** and select **WaterPumps**. For **Match Option**, use the drop-down arrow to select **Closest**. Keep all other options as default and click **OK**.

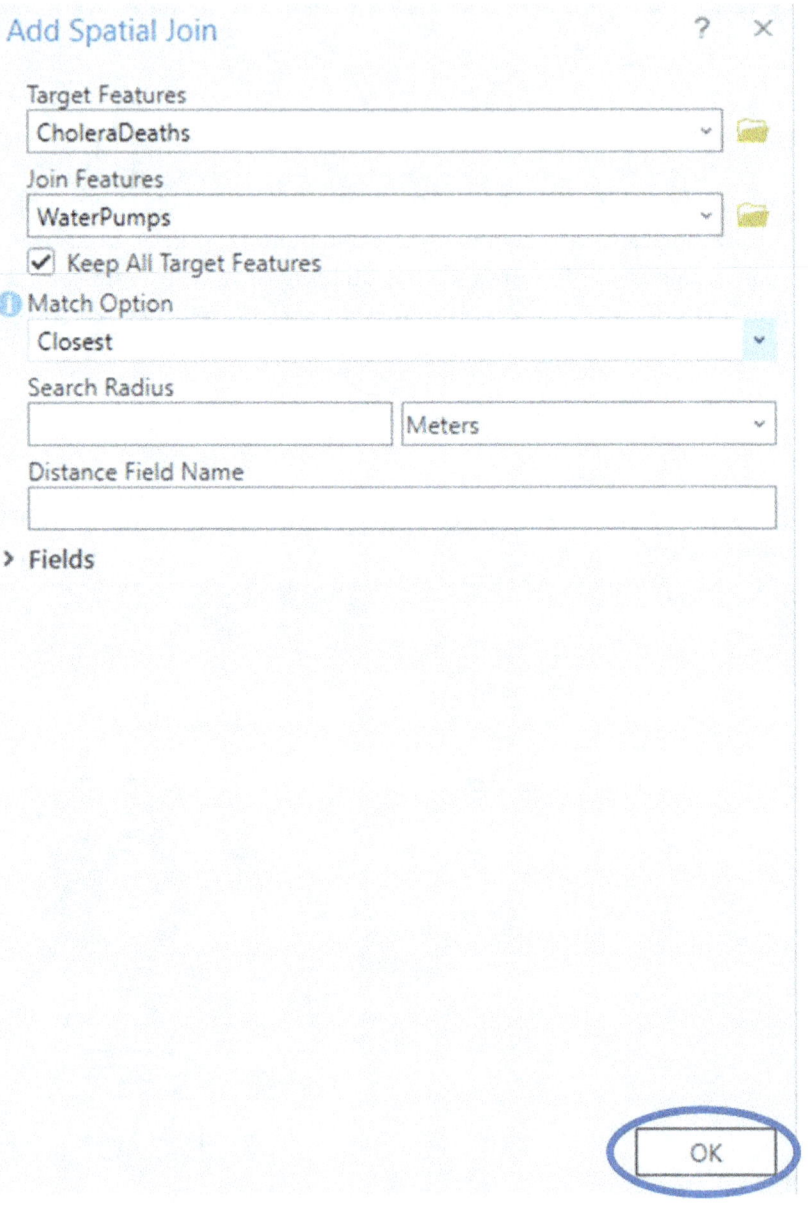

3. Once you have finished this, each feature in **CholeraDeaths** will be assigned the water pump that is within the closest distance to the point location.

Section 1.3 Task: Submit a screenshot of **CholeraDeaths** attribute table with the assigned features from **WaterPumps**.

Section 1.4: Creating a Buffer

In some cases, you want to know what study locations are located within a certain distance of a specific point. The **Buffer** tool will create polygons around a specific point, line, or polygon to a specified distance around the area. This will allow for easy extraction of study locations within these polygon buffers.

For this part of this tutorial, you will learn how to use the buffer tool.

1. Along the top of your ArcGIS Pro environment, click **Analysis**. Then choose **Pairwise Buffer**.

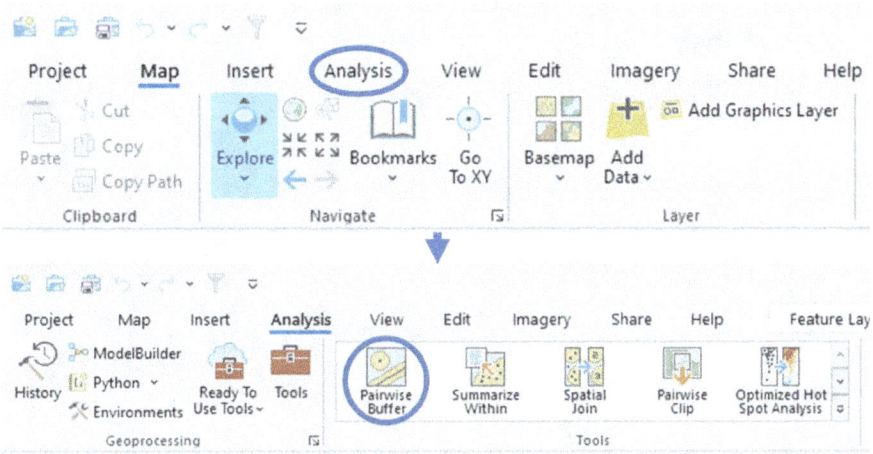

2. After clicking **Pairwise Buffer**, a pop-up window will appear at the right of your ArcGIS Pro environment. Click the drop-down arrow under **Input Features** and select **WaterPumps**. For **Output Feature Class**, click the folder next to the text box and navigate to the **JohnSnow.gdb** geodatabase. Name the new feature class **WaterPumps_Buffer** and click **Save**.

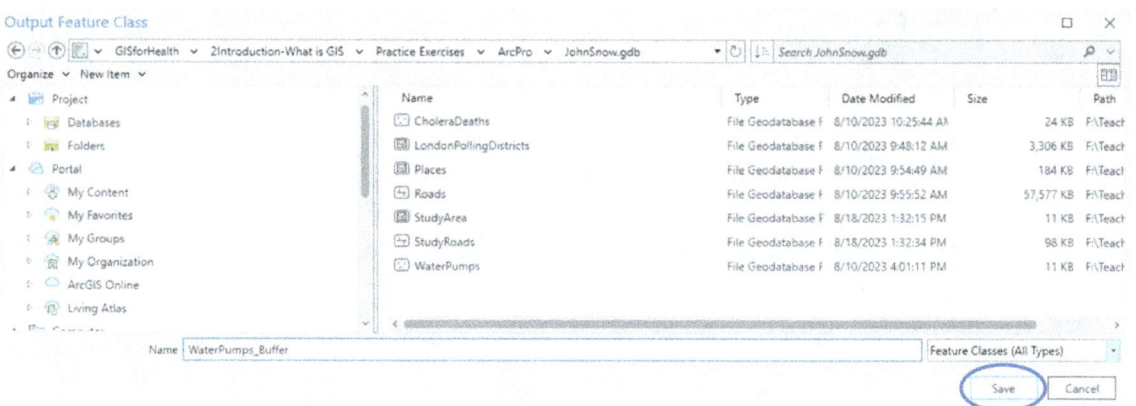

3. Continuing in the pop-up window on the right side of the ArcGIS Pro environment, there is a text box under **Distance**. In this text box, enter **250**. The input box next to the text box in which you entered 250, click the drop-down arrow and choose **US Survey Yards**. Keep all other options as default and click **Run**.

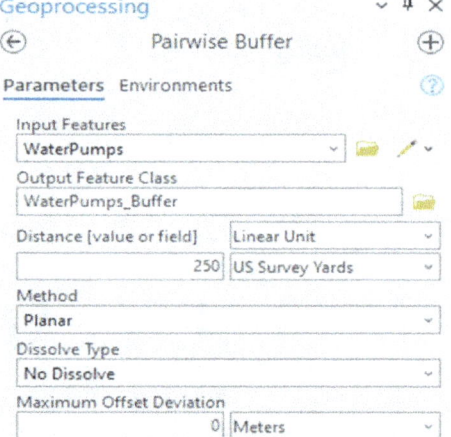

56

4. Once you have finished this, the new polygon feature called **WaterPumps_Buffer** will automatically be added to your map showing circles with a 250 yard radius around each water pump in the study area.

Section 1.4 Task: Submit a screenshot of your map **WaterPumps_Buffer** layer included.

Section 1.5: Changing Symbol Graphic Elements

Using the same symbol type for the same kind of features (i.e., circles for point locations) can make the map difficult to read, even if you use different colors. You can change the symbol's graphic element to make it easier to differentiate different types of features of the same kind.

For this part of this tutorial, you will learn how to change the graphic elements for point and line symbols.

1. In this exercise, we have point locations which include water pumps that service the study area (**WaterPumps**) along with point locations that symbolize where deaths occurred because of the cholera outbreak (**CholeraDeaths**). We want to change the graphic elements of the **WaterPumps** symbols to better identify the difference between these point locations. First, click the symbol below **WaterPumps** in the **Contents** pane on the left side of your ArcGIS Pro environment.

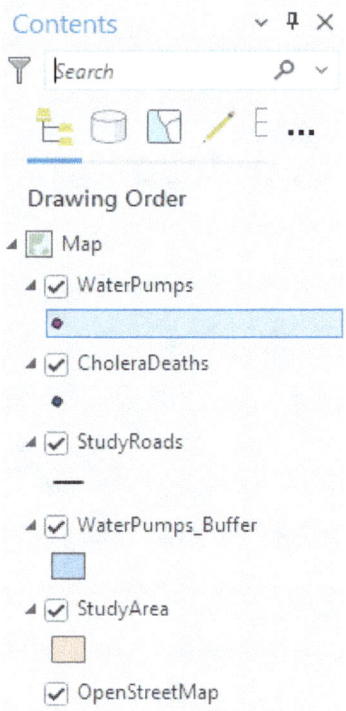

57

2. After clicking the symbol, a pop-up window will appear at the right of your ArcGIS Pro environment. In this pop-up window, click the symbol next to the word **Symbol**. In the search bar, type **water**. Search for the **Droplet** symbol under the **ArcGIS 2D** section and choose it.

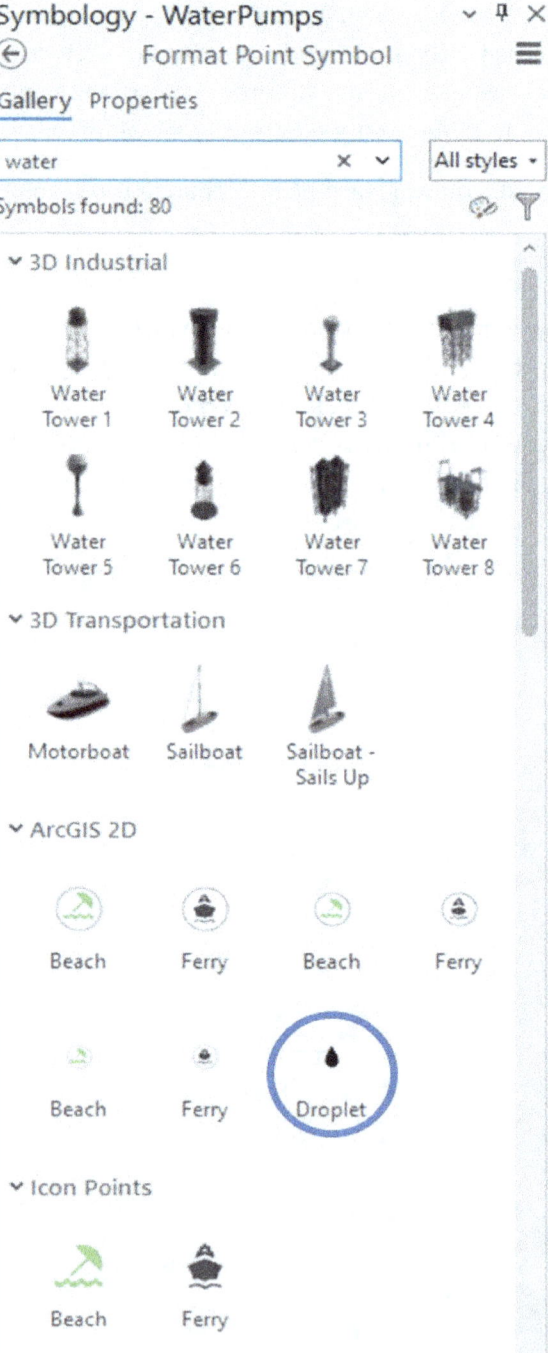

3. This changed all the features in the **WaterPump** layer to black water droplets. We want to change the color of these symbols to blue. In the **Symbology** pop-up window on the right side of the ArcGIS Pro environment, click **Properties**. Choose the drop-down arrow next to the color and select **Cretan Blue**. Click **Apply**.

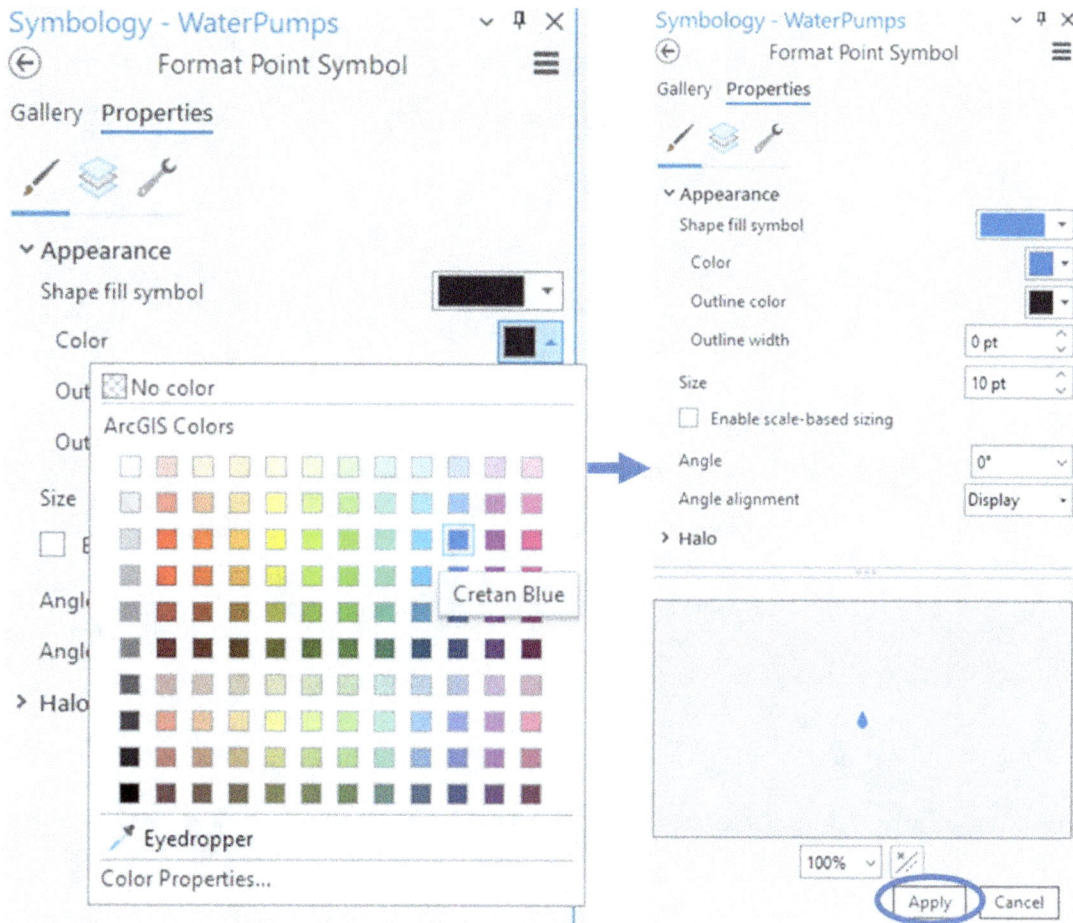

59

4. Changing the graphic elements for line features uses the same steps. First, click on the symbol below the **StudyRoads** layer in the **Contents** pane on the left side of the ArcGIS Pro environment. Then, click **Gallery** at the top the pop-up box on the right side of the ArcGIS Pro environment. Since these are just residential streets, choose the **Minor Road** symbol along the top line of choices.

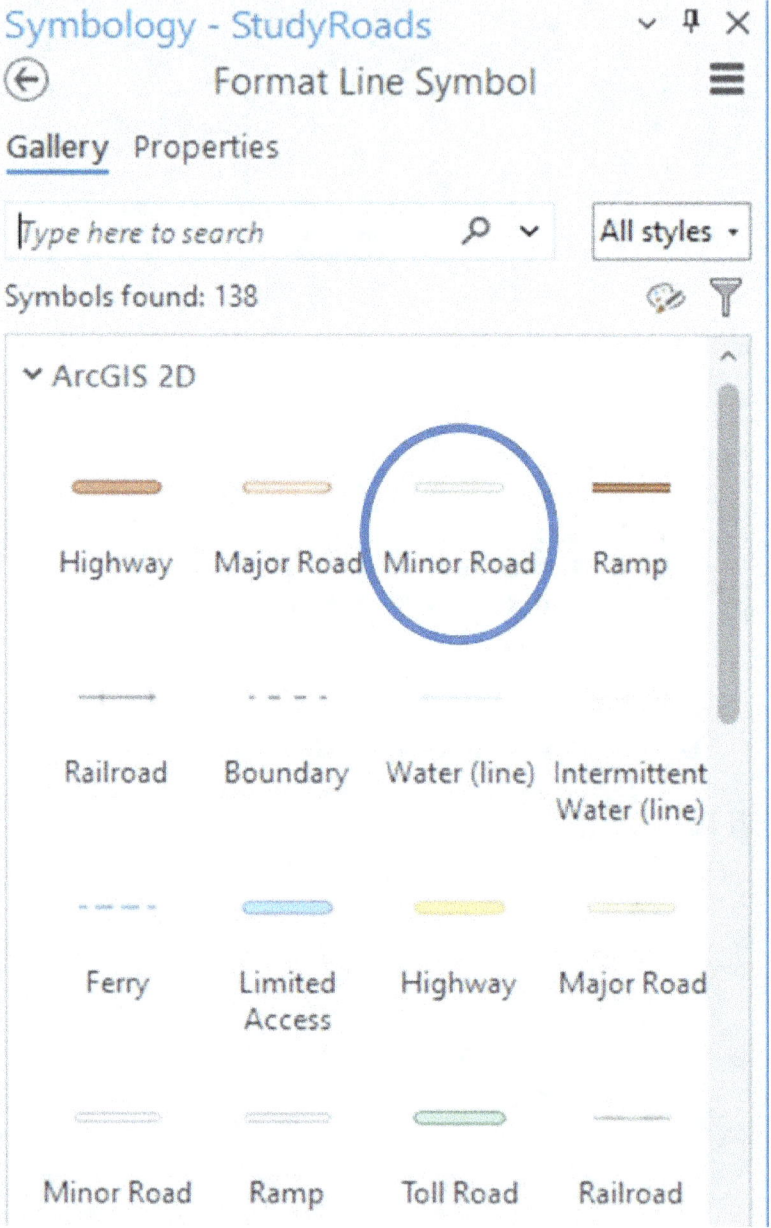

Section 1.5 Task: Submit a screenshot of your map showing the new symbols for the point and line features with the **WaterPumps_Buffer** layer deselected for better viewing.

Section 2: How to Visualize Spatial Data

Section 2.1: Graduated Symbols

When visualizing data, we can observe that some locations might have higher rates or frequencies than others, which can be stored in the attribute information. In our exercise, the **CholeraDeaths** point locations stores the number of cholera deaths at each point. Individuals can better visualize this information if we use graduated symbols. This will make areas with higher rates or frequencies have larger symbols.

This part of the tutorial will demonstrate how to create graduated symbols on your map.

1. Continuing in the **JohnSnow.aprx**, click the symbol under the **CholeraDeaths** layer in the **Contents** pane on the left side of the ArcGIS Pro environment. If needed, click the back arrow at the top the pop-up box on the right side of the ArcGIS Pro environment. Under **Primary symbology**, click the drop-down arrow next to **Single Symbol** and select **Graduated Symbols**.

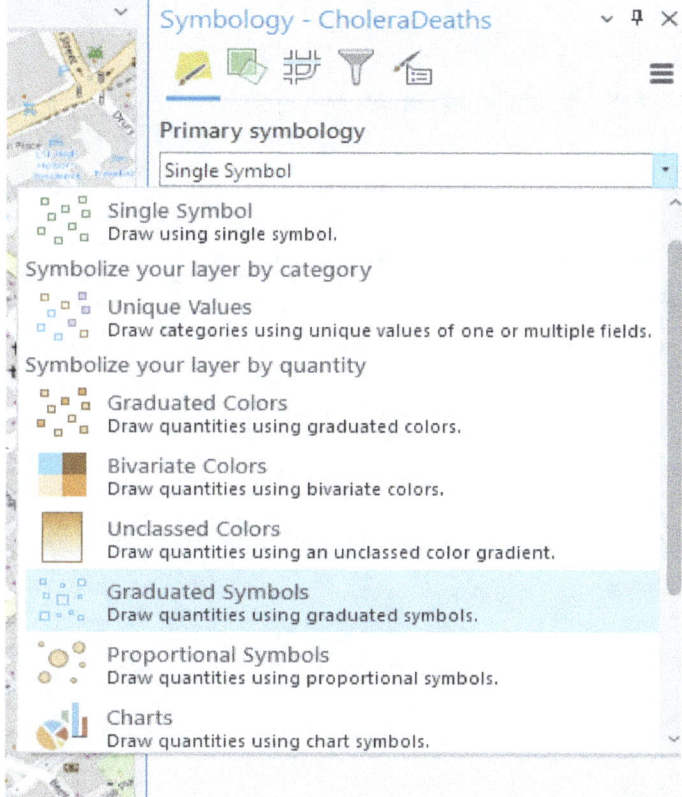

2. This changes the pop-up window to having a new form to fill out. Next to **Field** choose **DEATHS**. Next to **Template**, click the symbol. At the top of the pane, choose **Properties**. Click the drop-down arrow next to the **Color** and choose **Tuscan Red** and click **Apply**.

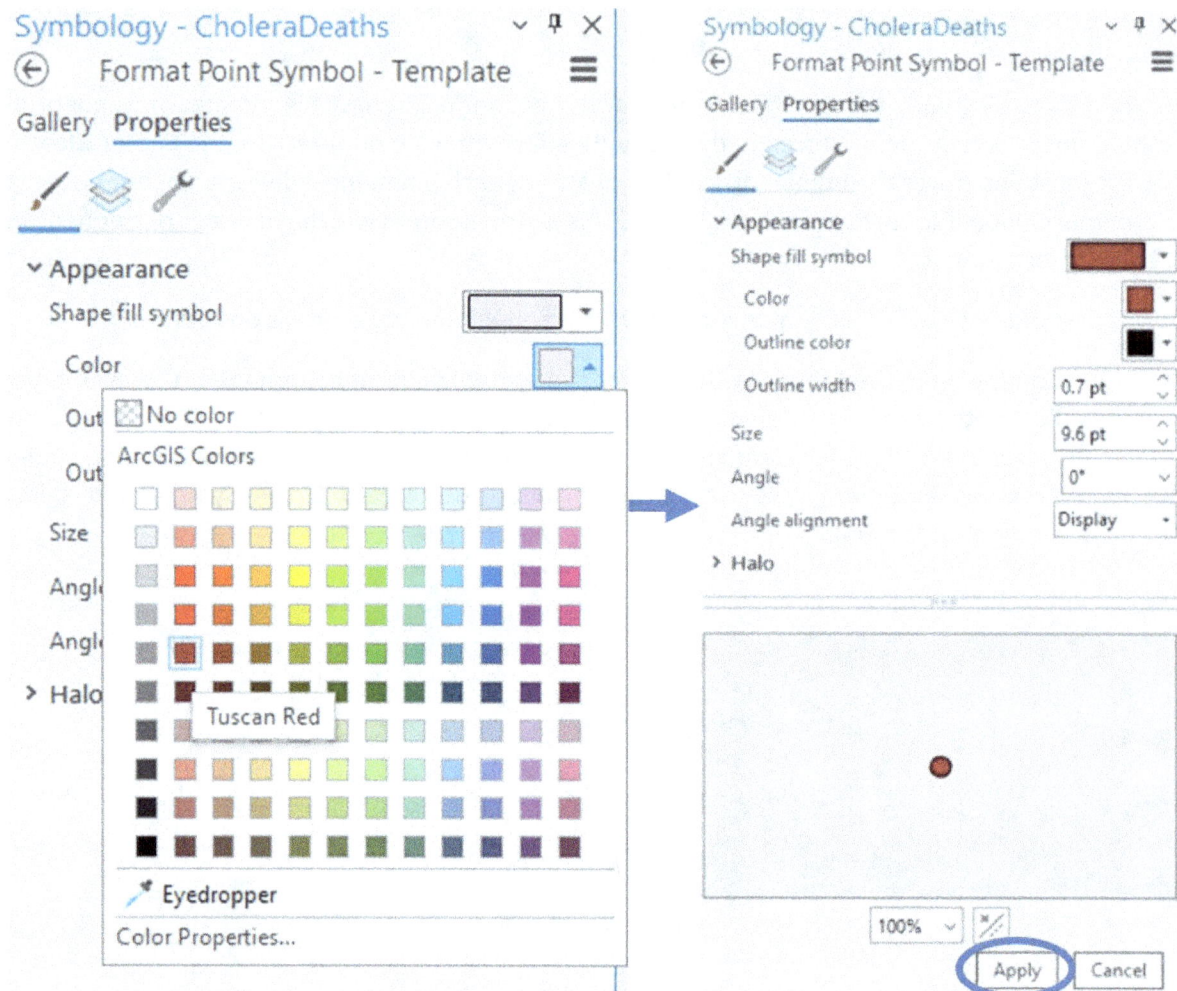

3. Click the back arrow at the top the pop-up box on the right side of the ArcGIS Pro environment. Keep all other information the same, and the form in the **Symbology** pane will now look like the figure below.

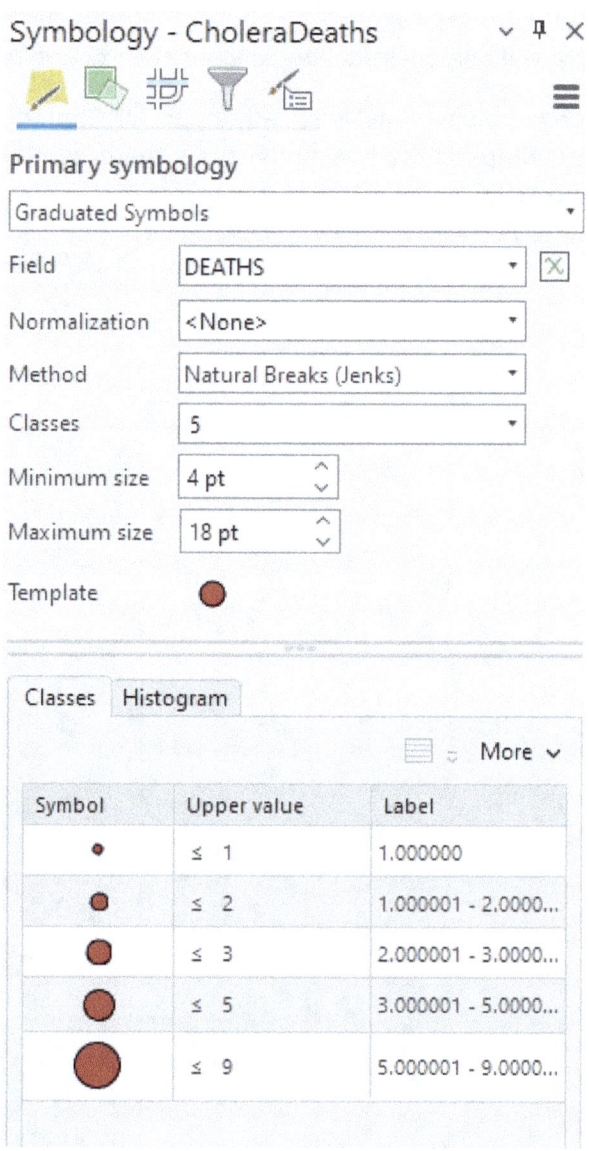

Section 2.1 Task: Submit a screenshot of your map on ArcGIS Pro with the graduated symbols.

Section 2.2: Choropleth maps

You can also view different rates or frequencies using varying shades of color. This is called a **Choropleth Map**. This is usually done with polygons, but in this case, we are going to do it for point locations. The steps to change the symbology are the basically same for both types of features.

This part of the tutorial will demonstrate how to create choropleth maps.

1. Because the **Symbology** pane is already open for **CholeraDeaths**, you can work from there. Click the drop-down arrow under the **Primary symbology** and choose **Graduated Colors**. Next to **Field** choose **DEATHS**. Next to **Color Scheme**, click the drop-down arrow and choose **Reds (5 classes)**.

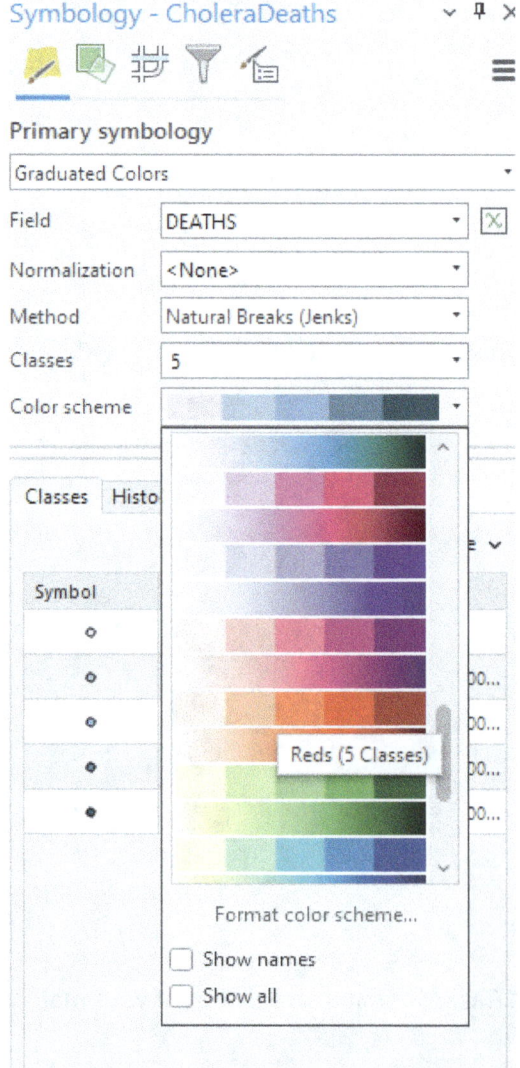

64

2. Keep all other information the same, and the form in the **Symbology** pane will now look like the figure below.

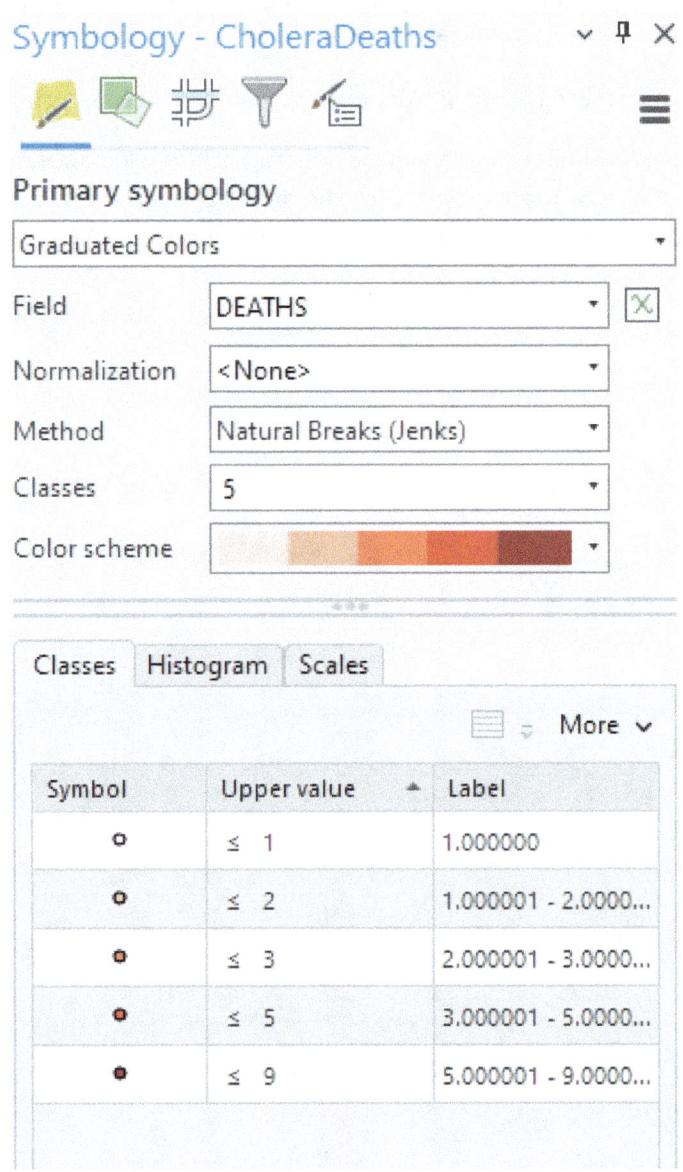

Section 2.2 Task: Submit a screenshot of your map on ArcGIS Pro with the graduated colors.

Section 2.3: Creating a heat map

We can use point layers to calculate relative density and display it easily on a map. This could be helpful to quickly visualize the highest rates or greatest frequencies in a specific area. Rates and frequencies can then be outlined in contours of different shading to demonstrate higher and lower occurrences. This is called a **Heat Map.**

This part of the tutorial will demonstrate how to create a heat map using the John Snow data.

3. This process is similar to changing symbologies. We will continue in the **Symbology** pane using the **CholeraDeaths** layer. Click the drop-down arrow under the **Primary symbology** and choose **Heat Map**. Next to **Weight Field** choose **DEATHS**. Keep everything else the same.

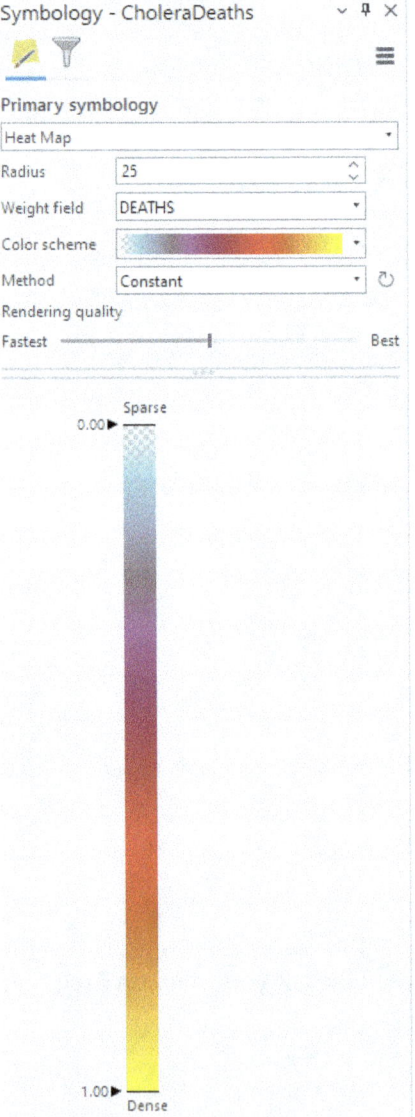

4. The result will show areas with the highest number of deaths in yellow, lowering into orange, then red, followed by purple, then blue, with areas with no deaths transparent. If you notice, there is one water pump in particular in the yellow area. If you click on this pump, it will tell you its name.

Section 2.3 Task: Submit a screenshot of your heat map on ArcGIS and answer the following question:

What is the name of the water pump in the yellow area of the heat map?

CHAPTER 3. CHRONIC DISEASES

Mastering the Concepts

This chapter examines the use of GIS in chronic disease research, emphasizing its role in identifying geographic disparities in diseases like cardiovascular diseases, stroke, cancer, respiratory diseases, obesity, and diabetes. As technological and social shifts have transitioned leading causes of death from infectious to chronic conditions, mapping these diseases has highlighted the influence of environmental factors in their development.

GIS tools enhance understanding by linking environmental variables—physical and sociocultural—to disease patterns, identifying high- and low-risk areas. They also support trend monitoring, intervention evaluation, and the design of effective chronic disease management programs, underscoring their transformative impact on public health.

Integrating disease registries with GIS allows health systems to analyze disparities in outcomes for chronic conditions such as diabetes, hypertension, and high cholesterol—key contributors to cardiovascular disease. Since socioeconomic status (SES) data is often absent from medical records, GIS integrates SES information from sources like the Census, enabling a deeper understanding of how demographic and neighborhood factors influence health outcomes. Using finer geographic units, like census tracts, enhances SES data accuracy, while longitudinal datasets enable tracking neighborhood changes and trends. This comprehensive approach supports policymakers in designing targeted strategies to reduce disparities and improve outcomes for patients with diabetes, hypertension, and high cholesterol.

SES variables aggregated at finer scales, like census tracts, better represent patients' conditions than larger, heterogeneous units. Longitudinal datasets allow trend analysis and tracking neighborhood changes over time. Additionally, using driving time as a more realistic measure of clinic accessibility, helping design cost-effective policies to address care disparities.

Mapping patient locations relative to primary care providers identifies service gaps, while analyzing regional and temporal trends, such as disease incidence and mortality, reveals patterns over time. Visualizing resources like support groups, community organizations or health services highlights areas for targeted interventions, such as education programs, preventive screenings, or expanded care. These insights are particularly valuable for addressing disparities in underserved or minority-dense areas, promoting equitable access

Geographic Information Systems (GIS) have played a crucial role in community-based health interventions, such as those aimed at addressing diabetes risk. By mapping and analyzing data related to health outcomes and social determinants, GIS helps identify areas with high health risks but low access to screening or treatment. This targeted approach allows initiatives to focus resources and interventions on communities where they are most needed. Through such strategies, GIS enables the identification of individuals at high risk or with undiagnosed conditions, ensuring that preventive measures and treatment referrals are directed efficiently to improve public health outcomes.

GIS plays a key role in obesity research by providing detailed spatial analysis of various factors influencing obesity. Using finer-scale data, such as student demographics, school districts, and zip codes, GIS can identify geographic patterns in obesity rates. These spatial tools allow researchers to map factors like BMI, physical activity, and socioeconomic status (SES) across different regions to guide interventions. For example, obesity rates are often mapped in relation to SES, ethnicity, and access to healthy food options, revealing areas with higher concentrations of overweight and obese populations.

GIS is used to analyze the built environment, including the proximity of fast-food outlets, grocery stores, and recreational facilities, and how access to food influences obesity rates. Neighborhoods with many fast-food outlets and convenience stores tend to have higher obesity rates, while areas with better access to grocery stores and fresh produce are generally associated with lower obesity risk. GIS also helps identify food deserts, areas where nutritious food is scarce, especially for those without cars. Figure 3.1 shows an example map of zip codes designated as food deserts.

FIGURE 3.1

Food desert areas by zip code in Shelby County, Tennessee, 2019.

Buffer analyses, which assess the distance to these resources, help identify "food deserts" and areas with limited physical activity opportunities. This geographic insight supports targeted policy and intervention efforts in communities with high obesity rates, particularly in underserved areas. Studies linking GIS data to environmental and socioeconomic variables further explore how factors such as residential density and neighborhood walkability contribute to obesity development.

To explore the relationship between obesity and environmental factors, studies rely on both primary and secondary data sources. Primary data includes surveys on environmental features, field observations, and data from technologies like GPS and remote sensing to analyze aspects like green space. Secondary data comes from a range of sources, such as census data, crime records, transportation networks, and business data. While primary data offers more specificity, secondary data is more cost-effective and can cover broader geographic areas. These data are often categorized into socioeconomic, physical, and personal attributes, which are key in understanding how the built environment influences obesity.

The U.S. Census and American Community Survey (ACS) provide key demographic data used in obesity research at the neighborhood level. These sources offer information on household income, education, and minority populations, which can be derived from 1-year, 3-year, and 5-year ACS estimates. Additionally, the National School Lunch Program serves as an indicator of low socioeconomic status (LSES), often coded by school and aggregated by higher geographic units like counties. These data help examine the links between sociodemographic factors and obesity.

Business data, often sourced from commercial databases like Info USA, are useful for neighborhood asset mapping, helping identify health-care facilities, food sources (e.g., fast-food restaurants), and recreational resources (e.g., parks, gyms). These datasets, including those from the NAICS, categorize facilities by type (e.g., eating places, outdoor venues). Police incident data help map crime areas that impact safety and restrict outdoor activities, while transportation infrastructure data, such as sidewalks, road crossings, and public transport access, are essential for examining the relationship between the built environment and physical activity. These data, often provided by local planning departments or transportation agencies, are crucial for spatial analysis in public health studies.

Personal environmental variables in obesity studies include self-reports on perceived environments (e.g., from children and parents), family history (e.g., diabetes, heart disease), and child fitness data (e.g., fitness levels of school children). These variables are collected through questionnaires and mapped at various geographic levels, such as zip code, school, or county. Perception data are particularly valuable in mixed-methods GIS studies on obesity.

GIS also plays a key role in cancer research by monitoring patterns, identifying health disparities, and guiding interventions. It helps focus cancer control efforts by providing georeferenced data on cancer rates at local levels, evaluating interventions, and predicting future trends. GIS is also used to identify cancer clusters, plan healthcare resources, and assess the impact of control efforts. The National Cancer Institute (NCI) has actively engaged in spatial data analysis and cancer mapping projects, such as developing web-based tools for cancer statistics (e.g., gis.cancer.gov). Through these initiatives, GIS enhances the understanding of cancer distribution and supports targeted health interventions.

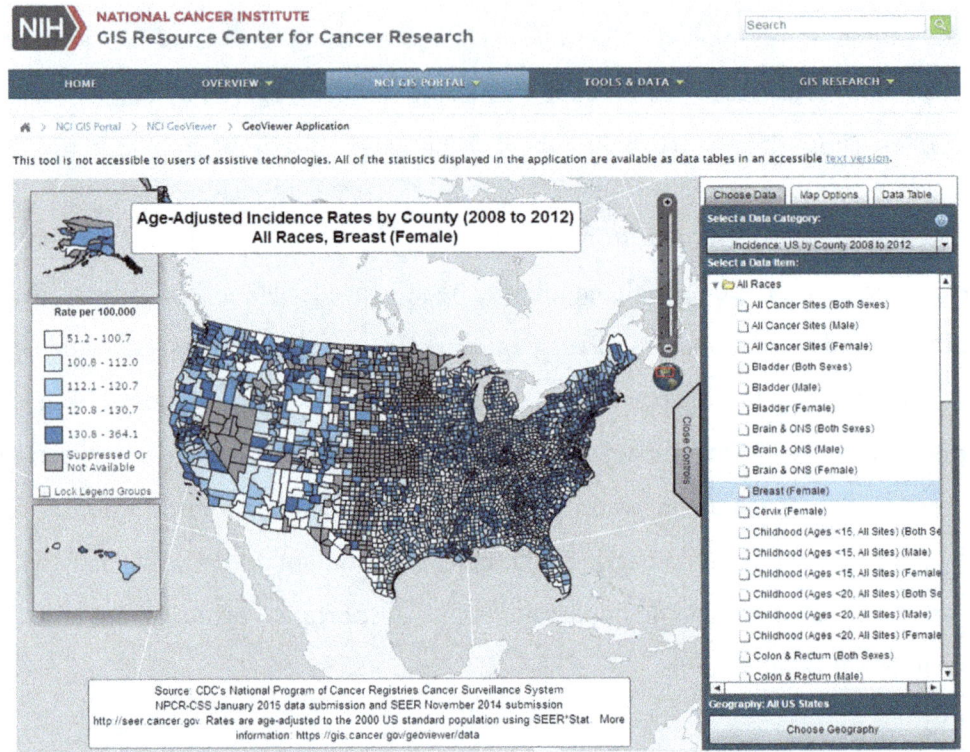

FIGURE 3.2 A screen shot from the National Cancer Institute GeoViewer Application. The map on the left has been created from the State Cancer Profiles data on the right.

Online, interactive health atlases, such as the State Cancer Profiles website, allow users to query and explore geographic variations in cancer data. These tools are useful for highlighting disparities, like higher cancer death rates and lower screening rates in specific populations, including those from low socioeconomic backgrounds, certain racial/ethnic groups, and isolated areas. GIS has yet to fully address these disparities, but its potential to track and analyze such health gaps is significant. Studies consistently show that African Americans and low-SES groups face more aggressive cancer diagnoses and poorer Cancer incidence and death rates in the U.S. show significant variations by race and ethnicity. Key modifiable risk factors, such as smoking, physical inactivity, and obesity, vary across these groups, with poor and minority communities often targeted by tobacco marketing and having limited access to healthy food and safe recreational spaces. Socioeconomic factors like income, education, and insurance coverage also influence access to early detection and treatment.

Geographic disparities in cancer types and burdens can be attributed to variations in risk factors, medical access, and cultural practices. In low-income countries, there is limited access to cancer treatment, especially in rural areas, due to a shortage of skilled professionals and resources. Cultural beliefs can further impact health behaviors and treatment choices.

Environmental factors play a significant role in cancer health disparities. Racial and ethnic minorities are disproportionately exposed to toxic environments, such as proximity to waste sites and high industrialization, increasing their cancer risk. GIS and spatial statistical methods are used to model exposure and analyze environmental justice, revealing that these communities often bear the brunt of environmental hazards, including toxic pollutants and pesticides.

With the growing global cancer burden, influenced by geography, ethnicity, and socioeconomic status, primary prevention through behavioral and environmental strategies is becoming crucial. Researchers are focusing on interventions, their impact on specific populations, and how to optimize their use. Identifying regions with high cancer burdens helps target areas for intervention. GIS mapping can uncover the combined effects of social, behavioral, environmental, and cultural factors, which may not be apparent in non-geographic studies. By linking population data to neighborhood attributes, GIS allows for investigating both individual and community influences on cancer patterns, even in the absence of individual-level data. GIS-based ecological studies examine factors like crime, green space, food deserts, and pollution that impact entire communities. However, these studies are typically cross-sectional, only identifying associations at the time of diagnosis without showing whether exposure occurred before, after, or during the onset of the disease. As a result, they cannot establish causality, although large sample sizes may help mitigate this issue.

GIS is essential for understanding the spatial patterns of infectious diseases and their link to chronic conditions. By identifying environmental and social factors that delay diagnosis and treatment, GIS provides insights into the spatial epidemiology of these diseases. This helps improve prevention, intervention, and access to healthcare, especially in areas with high health disparities. GIS also guides effective interventions by mapping disease spread and risk factors.

Mastering the Skills

Exercise 3: Using Health Data in ArcGIS Pro – Data cleaning, geocoding, and health data visualization

In this tutorial you are going to become familiar with adding health data to a map that does not have geographic coordinates. If the data provided has an address, you can use a **Geocoding** tool to match the addresses to geographic points in ArcGIS Pro. It then goes over some techniques for better visualizing the health data.

OBJECTIVES

- Clean up excel files with health information and upload into ArcGIS Pro
- Conduct address matching to geographic coordinate analysis (geocoding)
- Adding joins between shapefiles and excel files
- Create meaningful maps

Required data:

The original data sources listed below are provided for reference only. You do not need to download or curate these datasets from the original sources. All data, preconfigured GIS project files, and geodatabases required for this exercise are already included with the tutorial data and are available through the Resources page of the author's official website: www.esraozdenerl/resources. Instructions for accessing the tutorial data can be found in the Tutorial Data section of the Preface.

Original data sources:

1. Data from the Practice Exercises 3 folder. Sources:
    a. Health clinic locations (Excel): This publicly available data includes clinic locations practicing care for hypertension and participating in the Tennessee Heart Health Network (accessed on 09/14/2024).
    https://tnhearthealth.org/maps-portal/
    b. Hypertension data (Excel): PolicyMap. (accessed on 09/14/2023). Crude percent of high blood pressure among adults in 2019.
    https://policymap.com
    c. Minority census tract data (Excel):
        i. Source data is public HMDA Data for 2021 as of 2022-06-21.
        ii. Census tract minority population percentages are determined using the 2020 Census Redistricting Data, from the U.S. Census Bureau.
        iii. Tract to ZIP mapping is determined by Q4 2021 USPS ZIP Code Crosswalk data, as published by HUD.
        iv. 2020 block group to 2010 census tract mapping obtained from IPUMS National Historical Geographic Information System (NHGIS) at the University of Minnesota (https://www.nhgis.org/about-ipums-nhgis). For 2020 block groups spread across more than one 2010 census tract, we kept the pairing with the greatest overlap.
    d. City boundaries (Polygon): State of Tennessee Downloadable GIS Data
    https://tn-tnmap.opendata.arcgis.com/search?tags=boundaries
    e. Census tract boundaries (Polygon): U.S. Census Bureau
    https://myutk.maps.arcgis.com/home/item.html?id=ed71ab73dac14a93a7cd1020352bdfbc
    f. Zipcode boundaries and Shelby County boundary (Polygon): U.S. Census Bureau https://www.census.gov/geographies/mapping-files/2018/geo/carto-boundary-file.html

Section 1: Data Cleaning

Section 1.1: Clean up excel files with health information and upload into ArcGIS Pro

Researchers will get a lot of health information distributed to us in excel files. Sometimes this data needs minimal cleanup, and sometimes it can need a lot of cleaning in order to make it easily readable by ArcGIS Pro.

For the first part of this tutorial, you will learn how make excel files easily readable for ArcGIS Pro using three different excel files.

1. Navigate to the folder that includes the excel file data for Practice Exercise 3 and open **HealthClinicLocations.xlsx**. The file will look like the figure below.

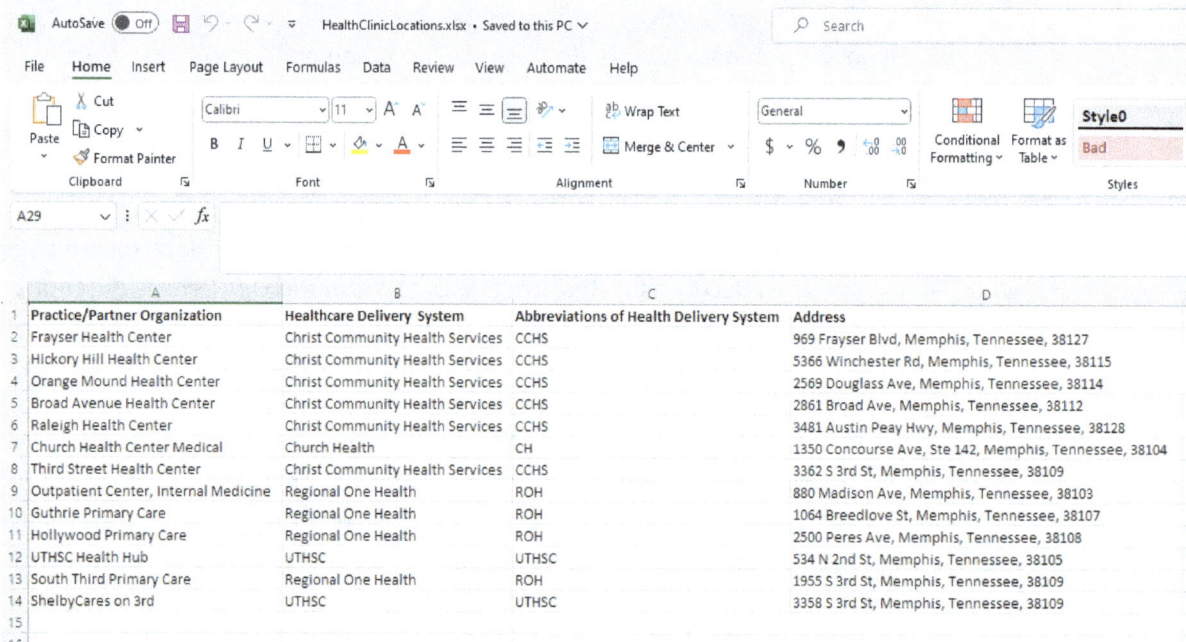

2. When trying to do address matching for later, you will want the address to be separated by street address, city, state, and zipcode. Therefore, we will have to split the last column into separate columns. Create the headers for each of the columns.

D	E	F	G	H
Address	Street Address	City	State	Zipcode
969 Frayser Blvd, Memphis, Tennessee, 38127				
5366 Winchester Rd, Memphis, Tennessee, 38115				
2569 Douglass Ave, Memphis, Tennessee, 38114				
2861 Broad Ave, Memphis, Tennessee, 38112				
3481 Austin Peay Hwy, Memphis, Tennessee, 38128				
1350 Concourse Ave, Ste 142, Memphis, Tennessee, 38104				
3362 S 3rd St, Memphis, Tennessee, 38109				
880 Madison Ave, Memphis, Tennessee, 38103				
1064 Breedlove St, Memphis, Tennessee, 38107				
2500 Peres Ave, Memphis, Tennessee, 38108				
534 N 2nd St, Memphis, Tennessee, 38105				
1955 S 3rd St, Memphis, Tennessee, 38109				
3358 S 3rd St, Memphis, Tennessee, 38109				

3. Fill in the appropriate information into each of the new columns. It will look like the figure below. **Note: To make things easier, you can copy and paste the Address into the Street Address column and delete after the street. All cities are Memphis, so you can type it in once and autofill. Same with all states: Tennessee. For the zipcode, you can copy and paste the address and delete everything before the zipcode.**

D	E	F	G	H
Address	Street Address	City	State	Zipcode
969 Frayser Blvd, Memphis, Tennessee, 38127	969 Frayser Blvd	Memphis	Tennessee	38127
5366 Winchester Rd, Memphis, Tennessee, 38115	5366 Winchester Rd	Memphis	Tennessee	38115
2569 Douglass Ave, Memphis, Tennessee, 38114	2569 Douglass Ave	Memphis	Tennessee	38114
2861 Broad Ave, Memphis, Tennessee, 38112	2861 Broad Ave	Memphis	Tennessee	38112
3481 Austin Peay Hwy, Memphis, Tennessee, 38128	3481 Austin Peay Hwy	Memphis	Tennessee	38128
1350 Concourse Ave, Ste 142, Memphis, Tennessee, 38104	1350 Concourse Ave, Ste 142	Memphis	Tennessee	38104
3362 S 3rd St, Memphis, Tennessee, 38109	3362 S 3rd St	Memphis	Tennessee	38109
880 Madison Ave, Memphis, Tennessee, 38103	880 Madison Ave	Memphis	Tennessee	38103
1064 Breedlove St, Memphis, Tennessee, 38107	1064 Breedlove St	Memphis	Tennessee	38107
2500 Peres Ave, Memphis, Tennessee, 38108	2500 Peres Ave	Memphis	Tennessee	38108
534 N 2nd St, Memphis, Tennessee, 38105	534 N 2nd St	Memphis	Tennessee	38105
1955 S 3rd St, Memphis, Tennessee, 38109	1955 S 3rd St	Memphis	Tennessee	38109
3358 S 3rd St, Memphis, Tennessee, 38109	3358 S 3rd St	Memphis	Tennessee	38109

4. When you finish, save the file by clicking **File** in the upper left of the Excel environment and selecting **Save**.

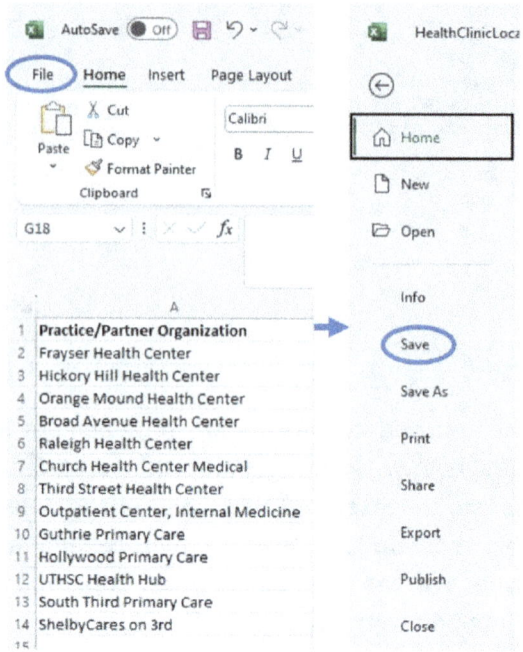

5. Now we have two more excel files to clean up. Next, open **HypertensionData2019.xlsx**. This file is mostly ready. However, you will notice that under state, the values are **Tennessee (State)**. We want this column to only say **Tennessee**. You can quickly fix this by conducting a replacement for that column. Click the drop-down arrow by **Find & Select** in the top panel of the Excel environment and choose **Replace**. Make sure to highlight the column for **GEOID** and check that the **Number** section says the column is **Text**. If not, use the drop-down arrow to choose that option. When you finish, save the file by clicking **File** in the upper left of the Excel environment and selecting **Save**. Just as before.

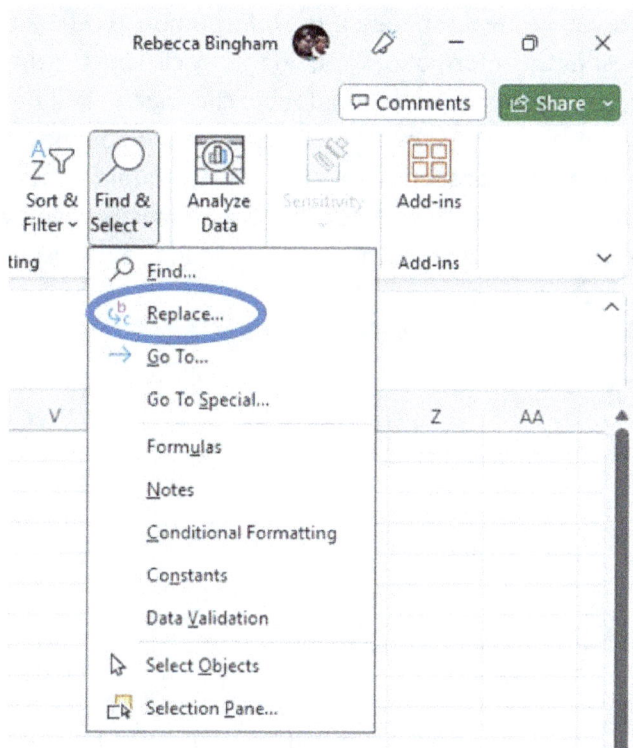

6. Lastly, we will clean up the **MinorityCensusTracts.xlsx** file. This file is a bit messy. First, you will want to delete the first three rows by clicking at the 1 on the left side of the excel environment and holding as you scroll down to the 3. This will highlight those rows. Then you right-click on the highlighted numbers and select **Delete**.

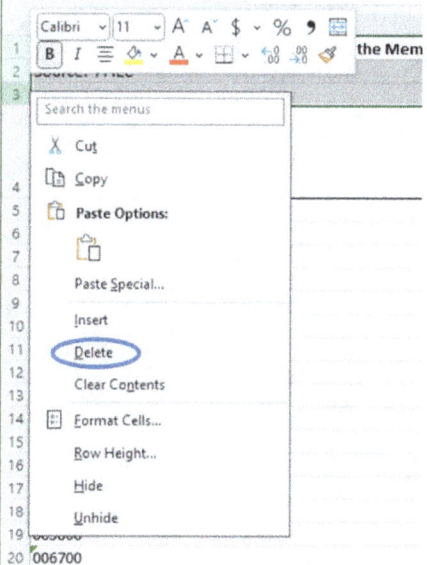

79

7. Next you will want to delete the information in **cell F1** and condense the columns to just fit the information. You do this by clicking the diagonal arrow in the top left corner to select all and then double clicking the line between columns A and B to condense them. Make sure to highlight the column for **Census Tract** and check that the **Number** section says the column is **Text**. If not, use the drop-down arrow to choose that option. When you finish, save the file by clicking **File** in the upper left of the Excel environment and selecting **Save**. Just as before.

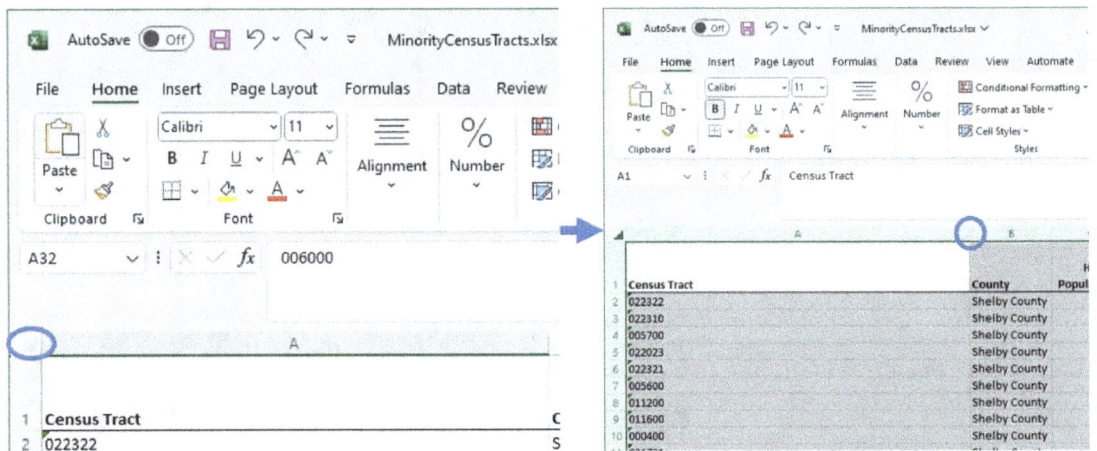

8. Now you are able to upload these excel files into ArcGIS Pro. You will want to open the **Hypertension.aprx** map file by navigating to that folder and double-clicking the file.

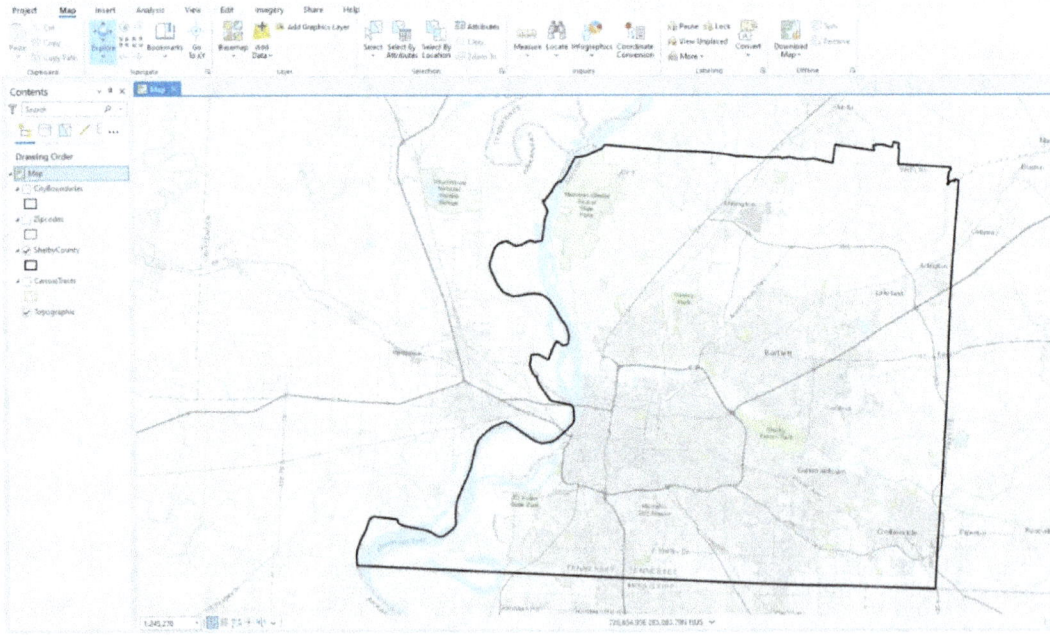

9. Next, you will have to set the data source for each layer in the map file as you did in Exercise 2. A short example of steps is included below. Do this for each layer.
   a. Right-click the **HealthClinics** layer->Choose Properties->Select Data->Click Set Data Source
   b. Navigate to the **Hypertension.gdb** geodatabase in the data folder for this exercise
   c. Choose the **HealthClinics** layer
   d. Click **OK**

10. Next, you will want to click **Add Data** in the top panel of the ArcGIS Pro environment. Navigate to your data folder with the saved excel files and add each of these files in. **Note: for each one, you will click on it and then only one sheet should be visible. Double click that sheet to upload it.**

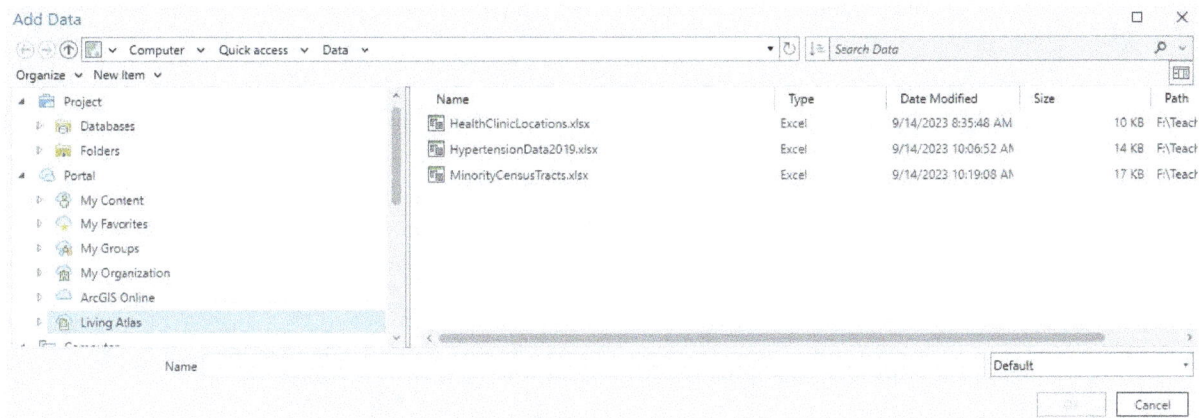

Section 1.1 Task: Submit a screenshot with all three cleaned-up data files loaded into ArcGIS Pro.

Section 2: Geocoding

Section 2.1: Conduct address matching to geographic coordinate analysis (geocoding)

Next, we must assign geospatial coordinates to our data in order for the data to be visible in the map on ArcGIS Pro. There are a couple of ways to be able to do this. The first is using the addresses in the health clinics file to be matched to geographic coordinates. This will create a point location file. The other is adding relates between an already established shapefile and the standalone tables.

This part of the tutorial will demonstrate how to geocode information that has addresses assigned to it.

1. Continuing within the **Hypertension.aprx**, right-click the **Locations$** table under the **Standalone Tables** section in the **Contents** panel on the left-hand side of the ArcGIS Pro environment and choose **Geocode Table**.

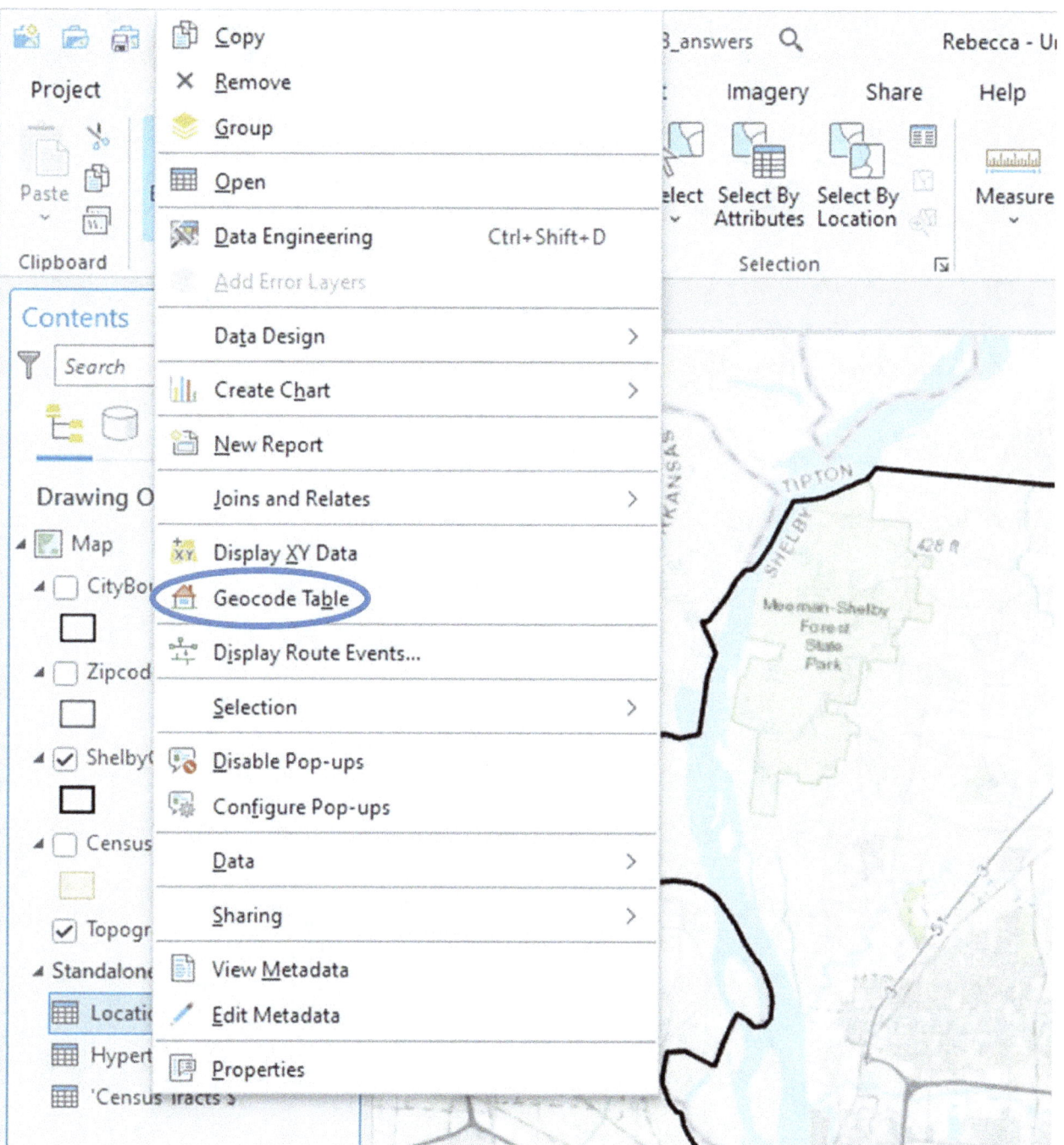

2. This changes the pop-up window on the right hand side of the ArcGIS Pro environment explaining the steps on how to geocode the data. Choose **Start** at the bottom right hand corner of this pop-up window.

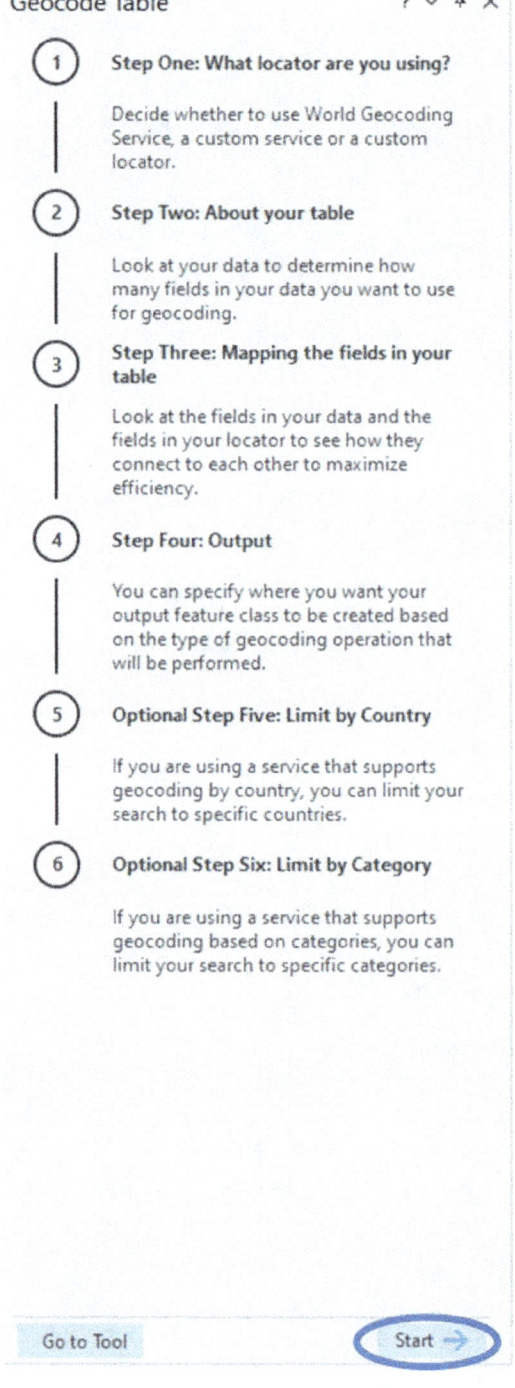

3. This brings you to **Step One**, where you choose the locator that you wish to use. Click the drop-down arrow next to the blank box under the **Input Locator** and choose **ArcGIS World Geocoding Service.** Then click **Next** on the bottom right-hand side of the pop-up window.

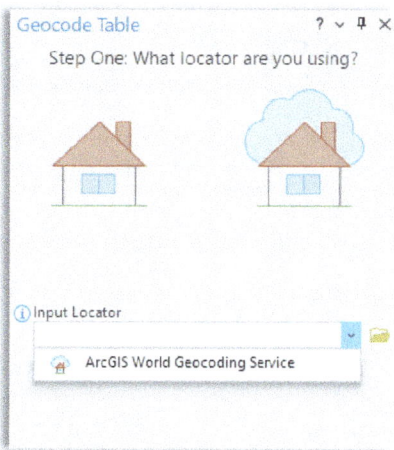

4. This brings you to **Step Two**, where you tell the locator about how the information is set up in your table. Here, you make sure that the **Input Table** says **Locations$**, and you verify that the next text box says **More than one field**. Then click **Next** on the bottom right-hand side of the pop-up window.

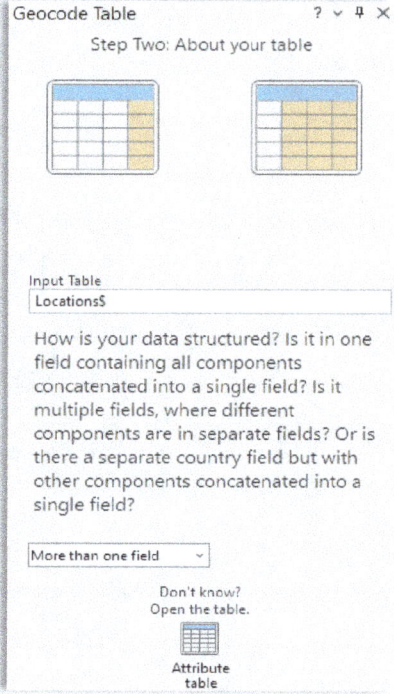

5. This brings you to **Step Three**, where you tell the locator the mapping fields in your table that relate back to information needed for geocoding. Here, you use the drop-down arrows by the data fields to make sure that the **Address or Place** says **Street_Address**, **City** says **City, State** says **State**, and **ZIP** says **Zipcode**. Then click **Next** on the bottom right-hand side of the pop-up window.

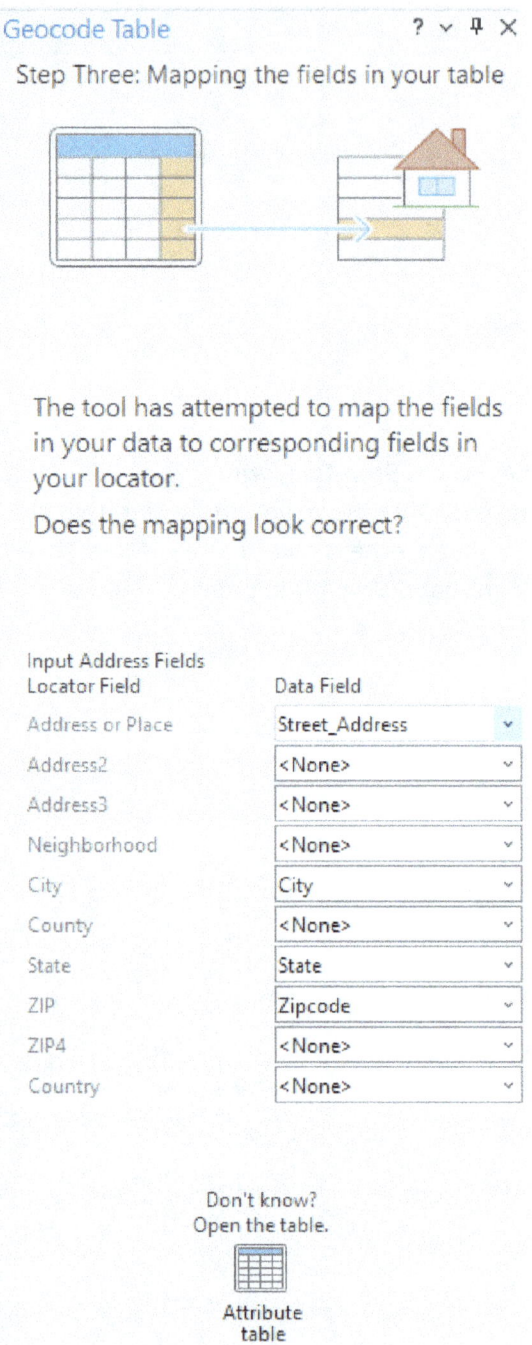

6. This brings you to **Step Four**, where you tell the tool where you want the data to be saved and what to name the new shapefile. Click the folder button next to the output field. Navigate to the **Hypertension.gdb** geodatabase folder and type **HealthClinics** next to **Name**. Select **Save**, then click **Next** on the bottom right-hand side of the pop-up window.

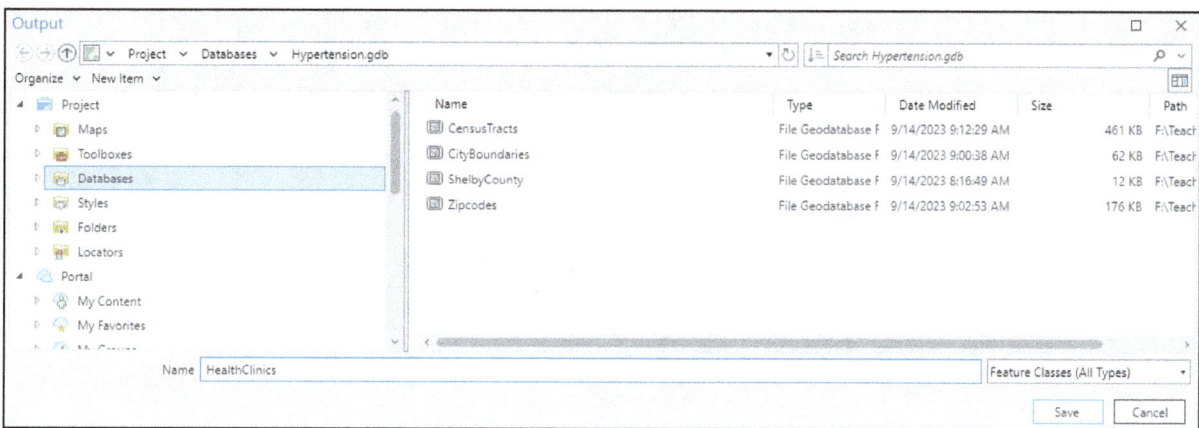

7. This brings you to **Step Five**, where you tell the tool what country(ies) in which your data was located. Click the checkbox next to **United States**. Then click **Next** on the bottom right-hand side of the pop-up window.

Geocode Table

Optional Step Five: Limit by Country

Is the data that you are geocoding located all over the world or in one or more countries?

Country
United States

- ☑ United States
- ☐ Afghanistan
- ☐ Albania
- ☐ Algeria
- ☐ American Samoa
- ☐ Andorra
- ☐ Angola
- ☐ Anguilla
- ☐ Antarctica
- ☐ Antigua and Barbuda
- ☐ Argentina
- ☐ Armenia
- ☐ Aruba
- ☐ Australia
- ☐ Austria

8. This brings you to the last step: **Step Six**. Here you tell the tool what categories you want to limit your address matching to. Click the checkboxes next to **Address** and **Postal**. Then click **Finish** on the bottom right-hand side of the pop-up window.

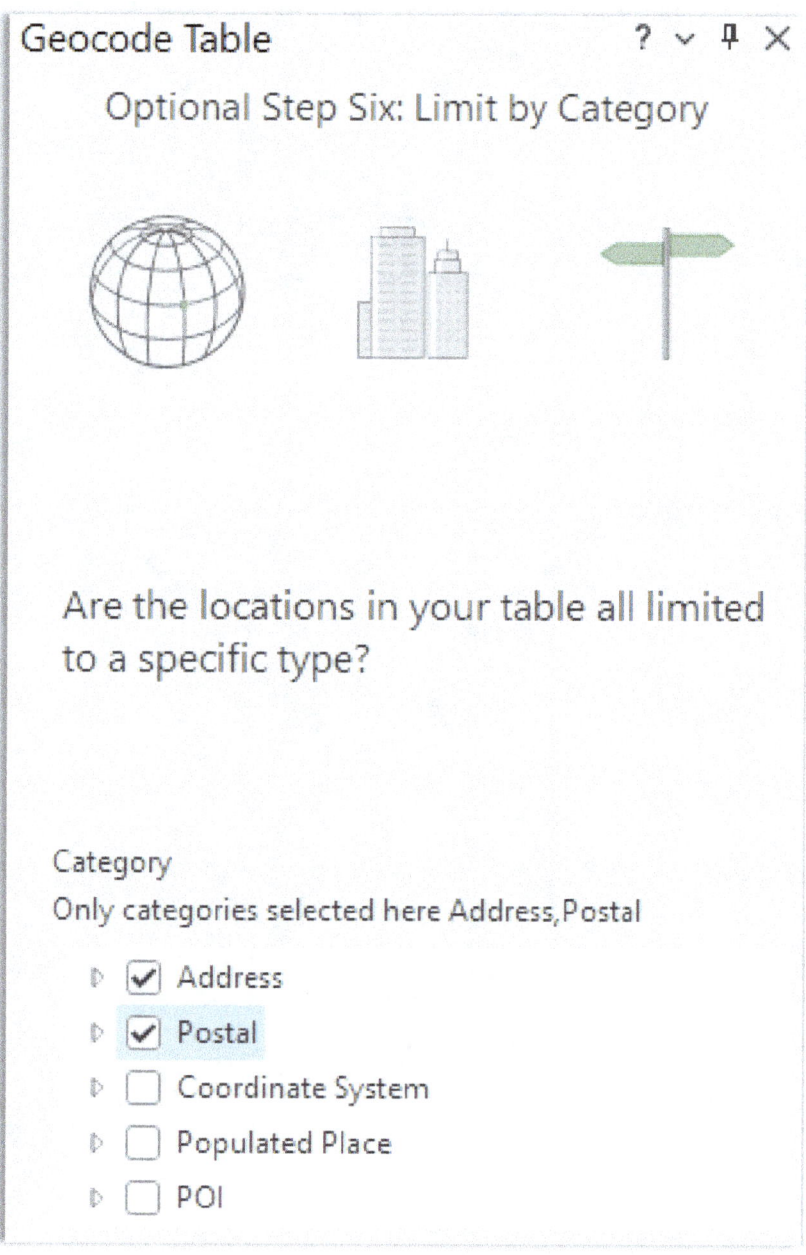

9. Finally, you want to click **estimate credits** in the top of the pop-up window and make sure all other information is filled out correctly. After this click **Run** on the bottom right-hand side of the pop-up window. This will add a new point shapefile with the health clinic locations onto your map. **Save** and **Close** your map project.

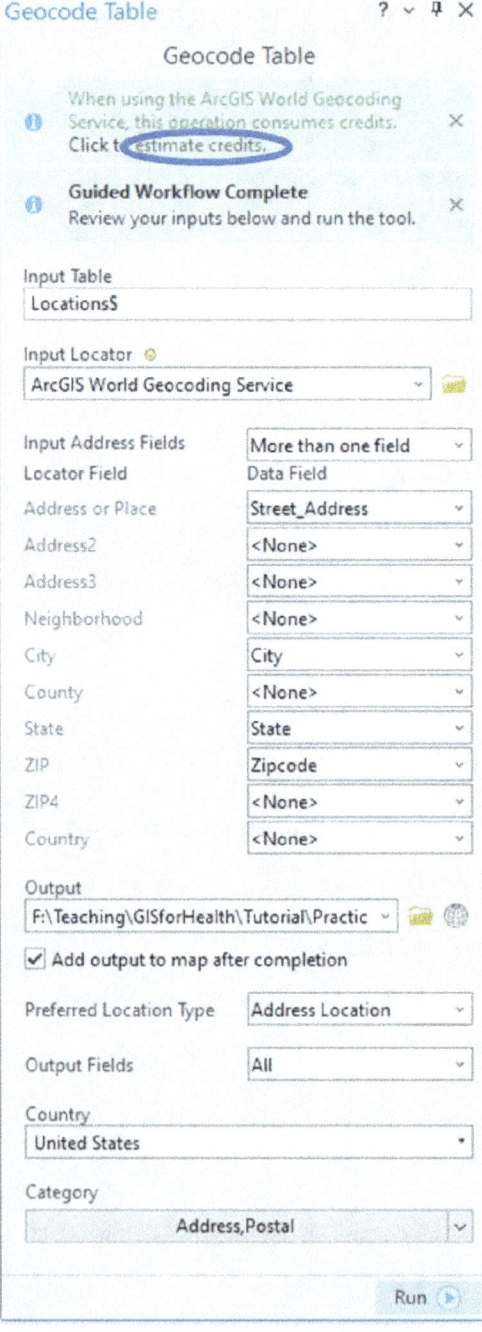

10. Now we want to clean up the attribute data for this new shapefile. Reopen the **Hypertension.aprx** map project. Then, open the attribute table for the **HealthClinics** shapefile. In the attribute table, you can see there is a lot of new information that is not needed. You can clean up this data by deleting some of these columns. First, switch to **Fields View** by clicking the three lines on the top right of the attribute table. To delete a column, right-click the box on the far left of the row select **Delete**. To delete multiple columns, use the **ctrl** button on your keyboard to select each of the columns you want to delete, then right-click one of the selected boxes on the far left of the row and select **Delete**. When you choose delete, the boxes on the far right will be highlighted green.

For this task, you will want to delete all columns except the following: **ObjectID ***, **Shape ***, **Status**, **Score**, **Match_type**, **Match_addr**, **X**, **Y**, **USER_Practice_Partner_Organization**, **USER_Healthcare_Delivery_System**, **USER_Abbreviations_of_HealthDelivery_System**, **USER_Address**, **USER_Street_Address**, **USER_City**, **USER_State**, and **USER_Zipcode**.

After choosing to delete all those fields, click the **X** next to **Fields: HealthClinics** to close the fields view. It will ask you if you would like to save changes. Click **Save**. When changes are completed and saved, the fields view window should close. If it does not, you tried to delete a column that cannot be deleted: **ObjectID ***, **Shape ***, **Status**, **Score**, **Match_type**, or **Match_addr**.

Section 2.1 Task: Submit a screenshot of your map on ArcGIS Pro with the health clinics point shapefile.

Section 2.2: Adding relates between shapefiles and excel files

Many times, you will want to take information from an excel file and add it as extra attributes to a shapefile. You can do this by adding a join between the shapefile and excel file. The result will be adding the information from the excel file as new columns (attributes) to the shapefile.

This part of the tutorial will demonstrate how to add joins between shapefiles and excel files.

1. There are two excel files that have information specific to certain census tracts: **Hypertension** and **'Census Tracts $'**. We want the data in these excel files to be added as additional information onto the census tracts shapefile. Therefore, we will be adding joins between the **CensusTracts** shapefile and those two excel files and creating a separate shapefile for each. First, you will need to right-click the **CensusTracts** shapefile, hover over **Joins and Relates**, and click **Add Join**.

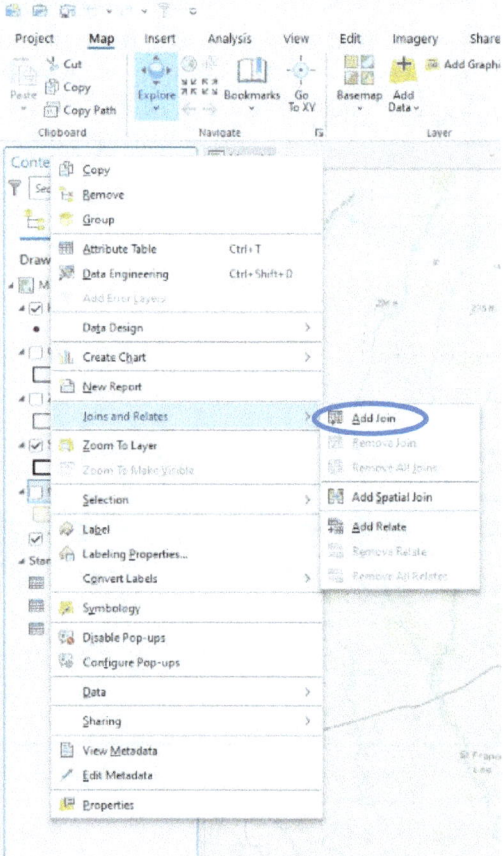

2. A pop-up window will appear. For the **Input Table**, use the drop-down arrow to choose **GEOID**. Under the **Join Table**, use the drop-down arrow to choose **Hypertension$**. For the **Join Table Field**, choose **GEOID**. Keep all other settings the same. Once the pop-up window looks like the figure below, click **OK**.

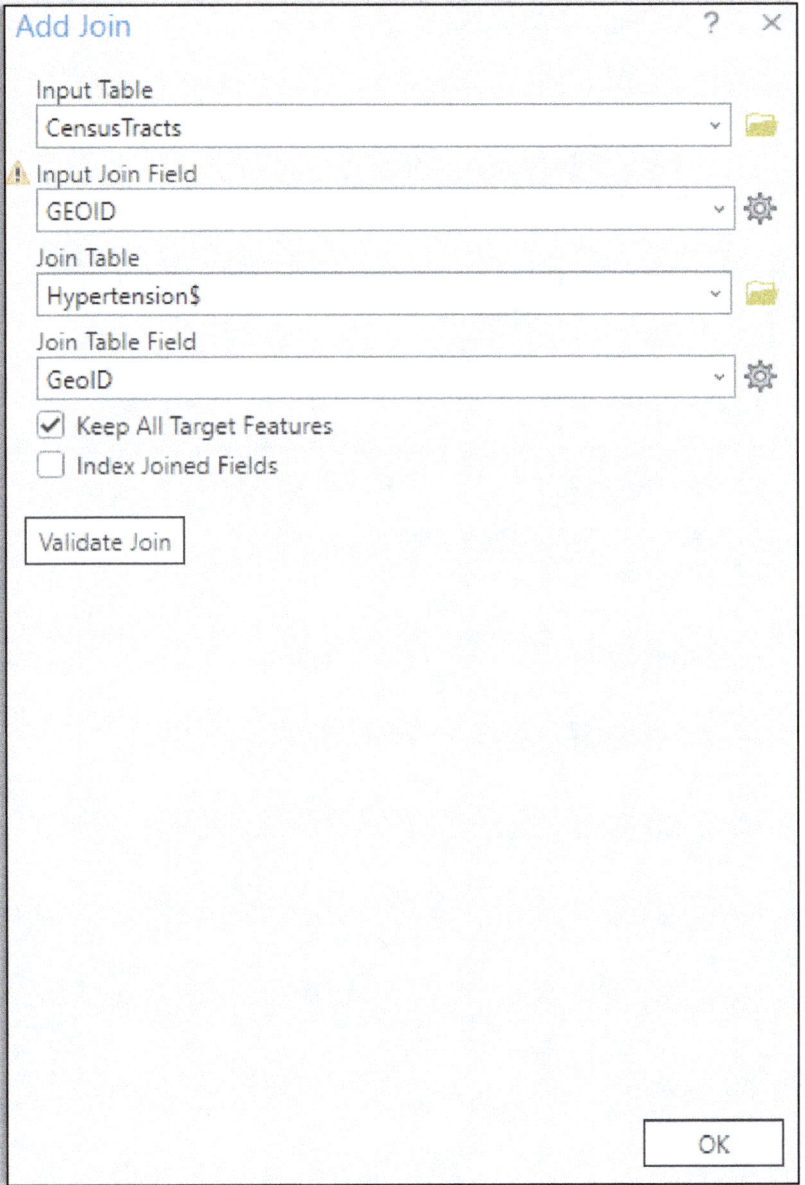

3. Next, you will want to extract this data and save it as a separate shapefile with the hypertension data. Right-click **CensusTracts**, hover over **Data**, choose **Export Features**.

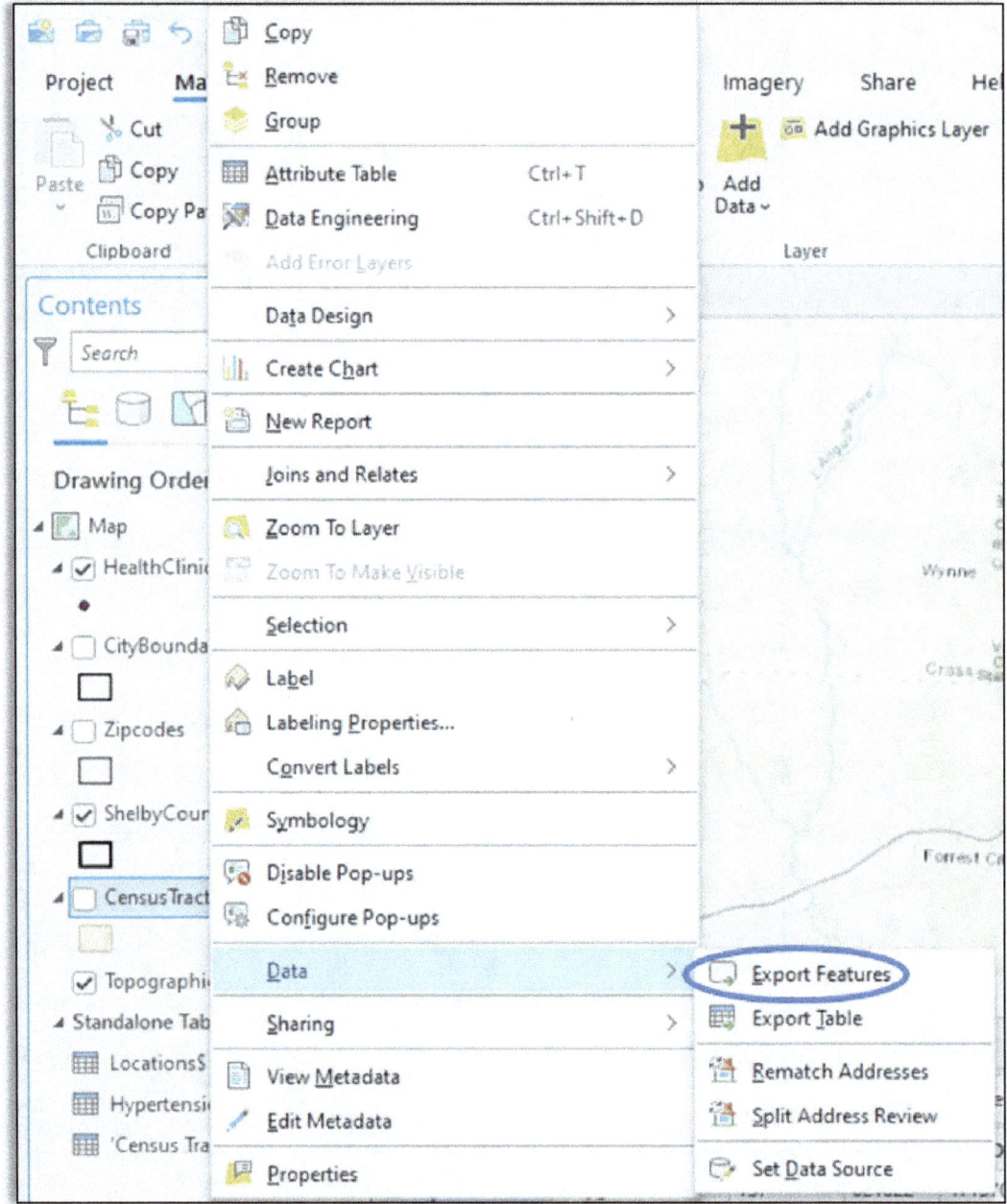

4. A pop-up window will appear. Click the folder next to the text box under **Output Feature Class**. Navigate to the **Hypertension.gdb** geodatabase. Name it **CensusTracts_HighHypertension**. Click **Save**, then **OK**. If you look at the attribute table, you will notice that there are many census tracts with null values for the new data. This is because the **Hypertension$** excel file only had information for those census tracts whose population had high rates of hypertension.

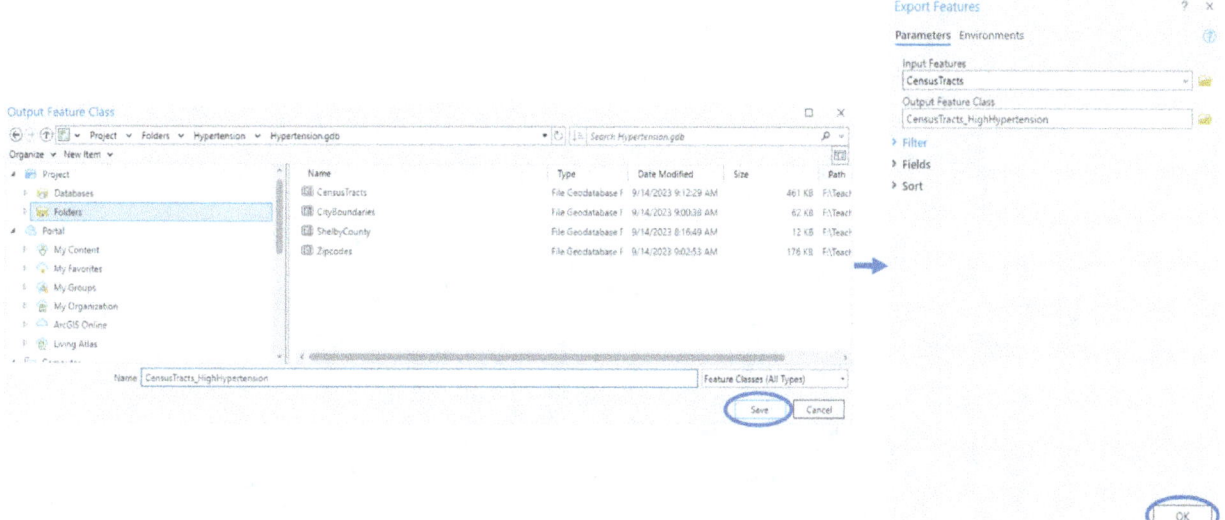

5. Now we will want to add the '**Census Tracts $**' attribute information to the **CensusTracts** shapefile. First, you will need to remove any previous joins. Right-click the **CensusTracts** shapefile, hover over **Joins and Relates**, and click **Remove All Joins**. Then click **Yes**.

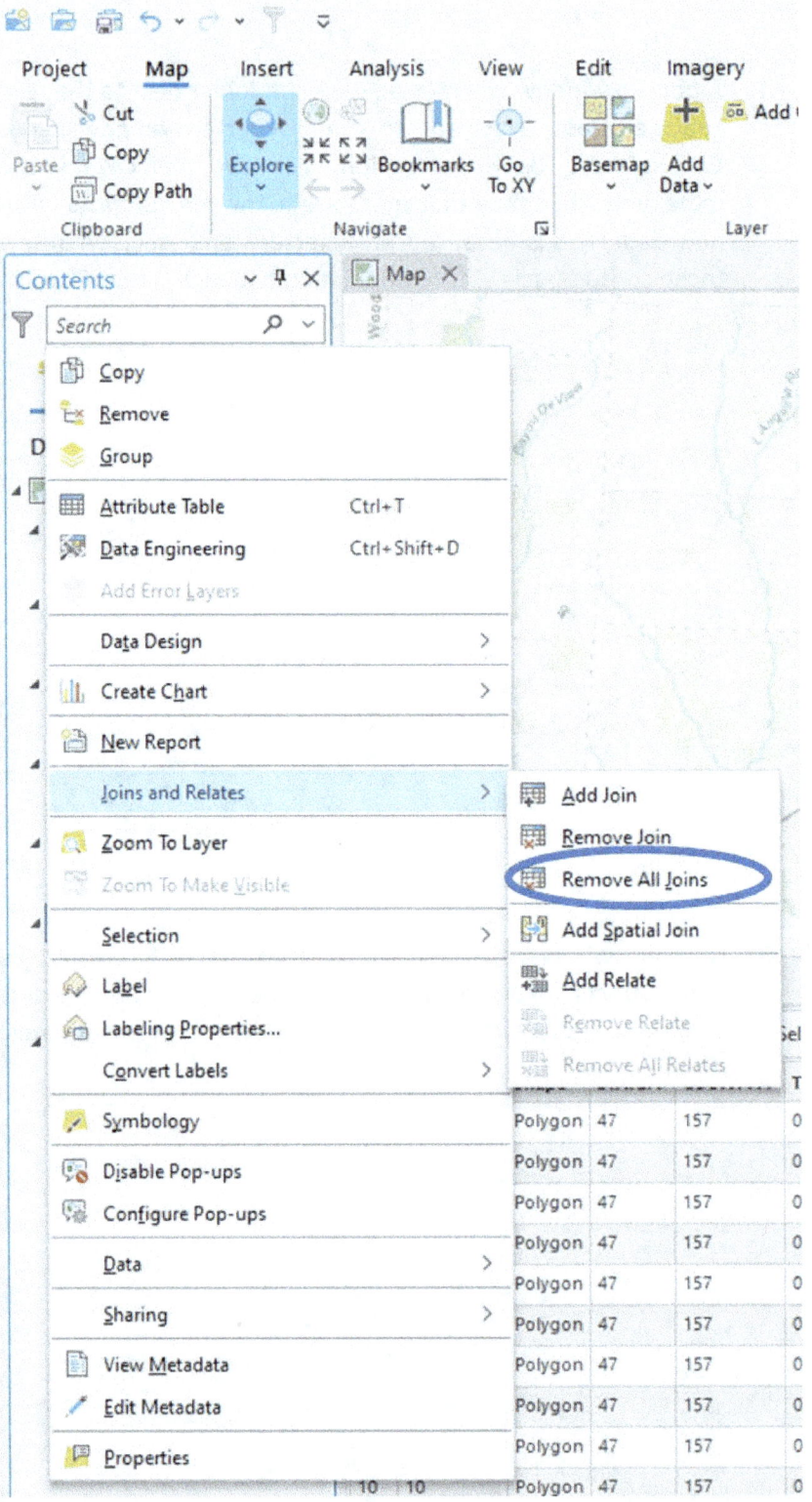

96

6. Then you can repeat the previous steps to add a join between the **CensusTracts** shapefile and the **'Census Tracts $'** excel file. Once the pop-up window looks like the figure below, click **OK**.

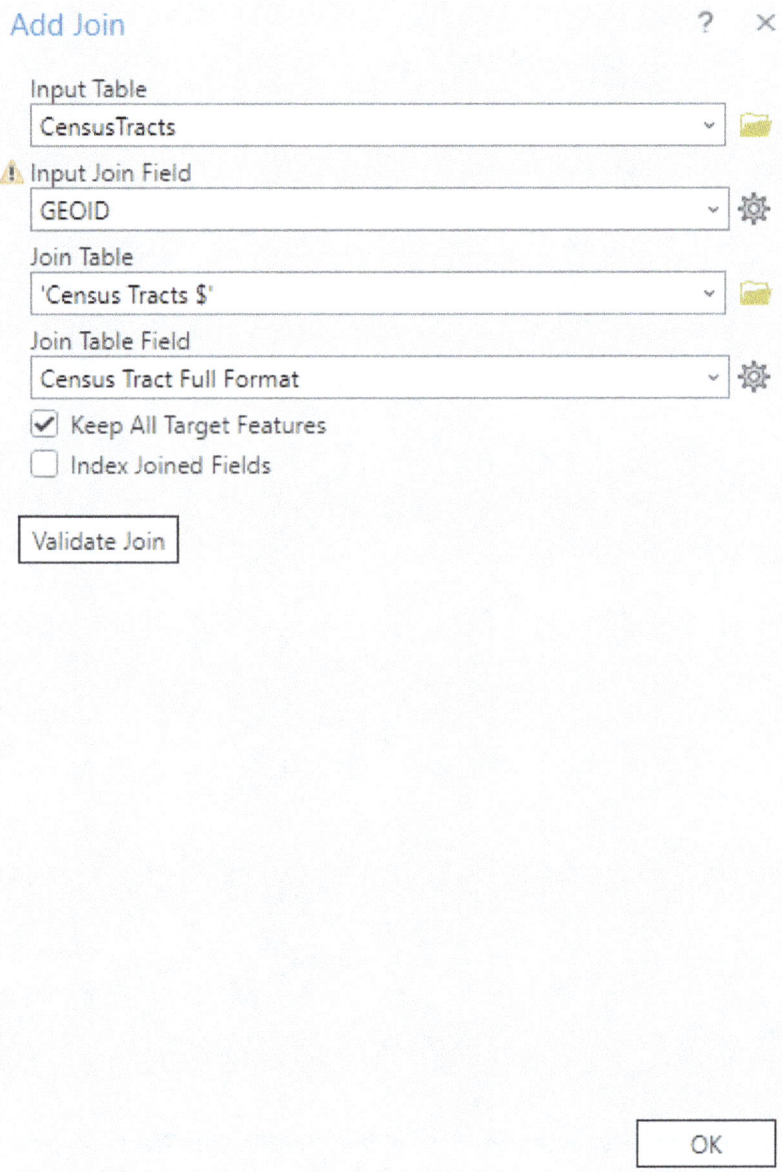

7. Next, you will want to extract this data and save it as a separate shapefile with Black and Hispanic population data. You will want to save it as **CensusTracts_MajorityBlack_Hispanic**. See the figure below. Remember to remove all joins once you have exported the data into a new shapefile.

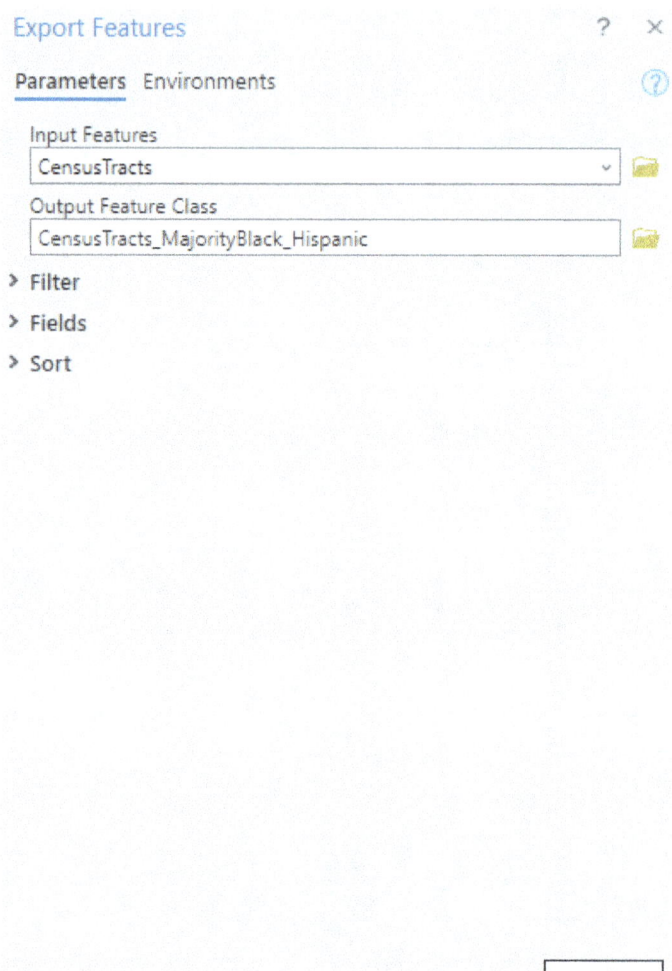

8. Like the **Hypertension$** excel file, the '**Census Tracts $**' excel file only had census tracts whose populations were comprised of majority Black or Hispanic individuals. This leaves a lot of census tracts with null values for the new attributes added to it. Keep this in mind as you move down to future tasks.

Section 2.2 Task: Submit a screenshot of your map on ArcGIS Pro with two new shapefiles: **CensusTracts_MajorityBlack_Hispanic** and **CensusTracts_HighHypertension**, along with the attribute table for **CensusTracts_MajorityBlack_Hispanic** open to show the data for it.

Section 3: Health Data Visualization

Section 3.1: Create meaningful maps

Right now, you have all the data in your map, but it still doesn't really tell a story. To make a good health data map, you have to present the data in a way that makes it easy for people to quickly understand what you are trying to tell them.

For the final part of this exercise, you will learn how to create meaningful maps.

1. Let's begin by looking at areas with high hypertension. For this map, you will want to show the viewers where there are areas of high hypertension and how close the health clinics that service people with high hypertension are located. First, turn off any layer not needed to show this information:
CensusTracts_MajorityBlack_Hispanic.

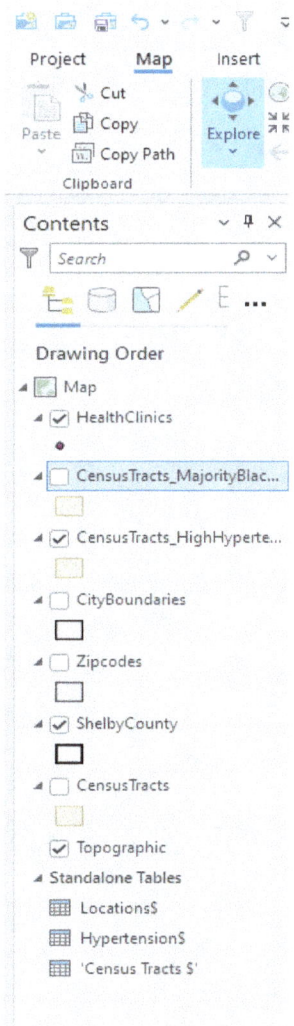

2. This still does not really show what we are looking for. We need to separate census tracts that have high rates of hypertension from those that do not. Remembering from earlier, the **CensusTracts_HighHypertension** layer has null values in the newly provided attribute information for counties that do not have high rates of hypertension. We can use this to help create a symbology that separates them from the rest.

   Right-click **CensusTracts_HighHypertension** and choose **Symbology**.

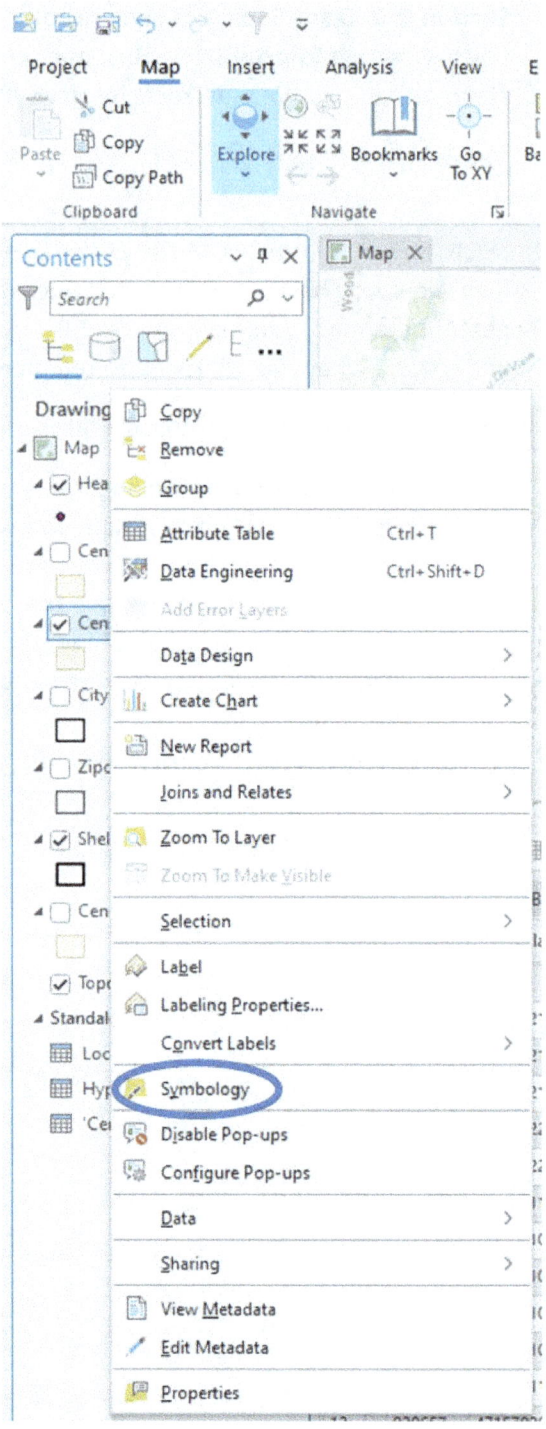

3. We are also going to want to see the city boundaries for this map, so we want to turn on the **CityBoundaries** layer and move it to the top of the map. Turn on the labels for this layer by right-clicking the layer and selecting **Label**. Notice that the only health clinic data we have is for the city of Memphis. This is going to be our target location. You will want to zoom into this area. First, choose **Select By Attributes** at the top of the ArcGIS Pro environment. Next, make a selection for those cities whose name is Memphis by making sure that the **Input Rows** has **CityBoundaries** selected, the **Where** field has **NAME** selected, and Memphis is in the final dropdown menu box. Hit **Apply** and **OK**. Then make a layer from that selection by right-clicking the layer, hovering over **Selection**, and choosing **Zoom To Selection**.

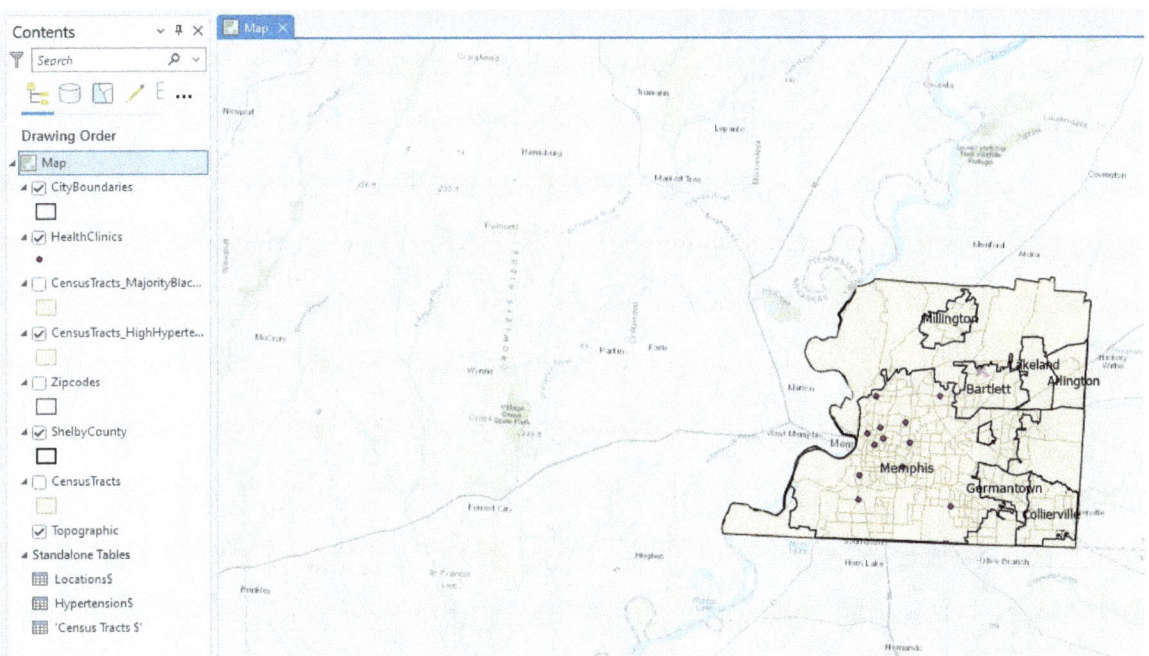

4. Notice that the only health clinic data we have is for the city of Memphis. This is going to be our target location. You will want to isolate this city from the rest. First, choose **Select By Attributes** at the top of the ArcGIS Pro environment. Next, make a selection for those cities whose name is Memphis by making sure that the **Input Rows** has **CityBoundaries** selected, the **Where** field has **NAME** selected, and Memphis is in the final dropdown menu box. Hit **Apply** and **OK**.

5. Next, you will make a layer from that selection by right-clicking the layer, hovering over Selection, and choosing **Make Layer From Selected Features**. Make this layer permanent by right clicking **CityBoundaries Selection**, hovering over **Data**, and choosing **Export**

Features. Click the folder next to the textbox for **Output Feature Class**. Navigate to the **Hypertension.gdb** geodatabase and name the layer **Memphis**.

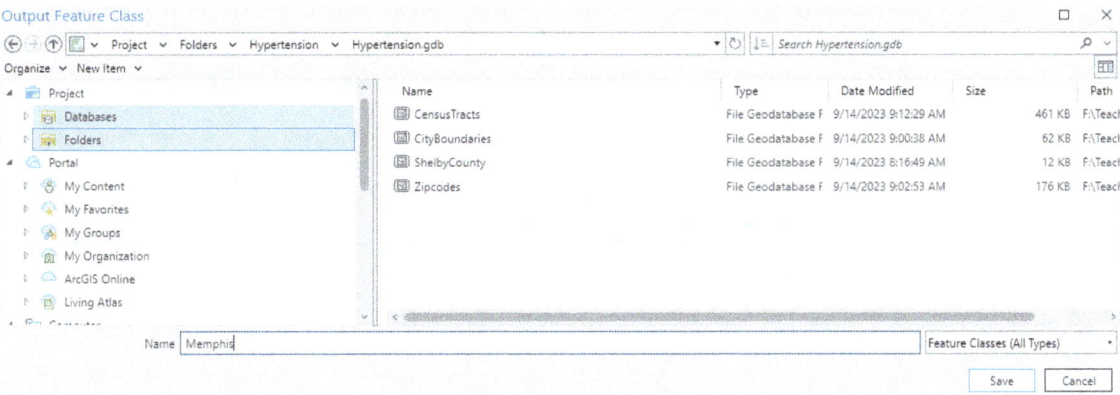

6. Click **Clear** at the top of the ArcGIS Pro environment to clear the selected features, remove the **CityBoundaries Selection** layer, and turn off the **CityBoundaries** layer. Then, right-click the **Memphis** layer and choose **Zoom To Layer**.

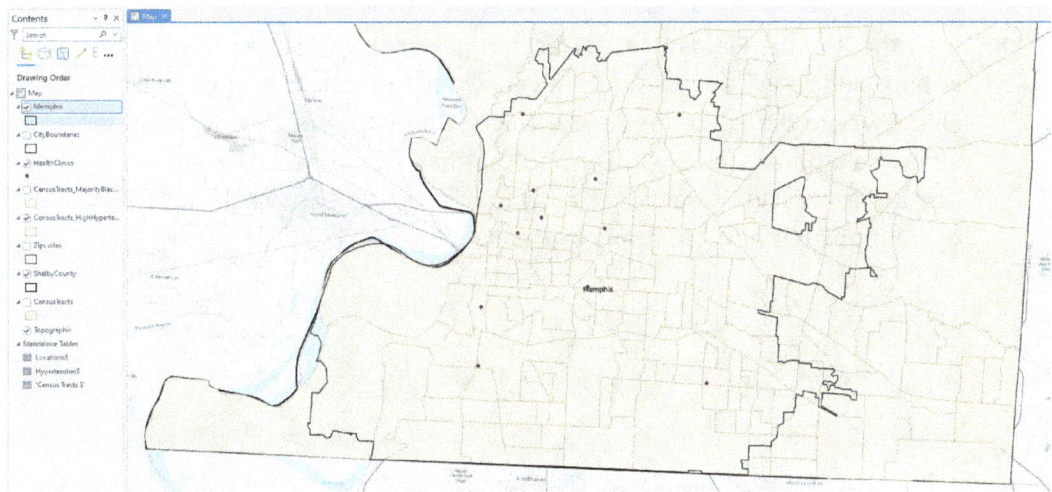

7. Next, we want to be able to visualize areas with high hypertension rates. To do this we can create a new field that is called **Hypertension** and assign areas that have high hypertension rates **High** and **Not High**. First, open the attribute table for **CensusTracts_HighHypertension**. Then, click **Add** at the top of the attribute table. This will open the **Fields View** with a new field on the bottom row for you to fill out. For **Field Name** enter **Hypertension** for **Alias** enter **Hypertension**. For **Data Type**, choose **Text**. Leave the rest alone.

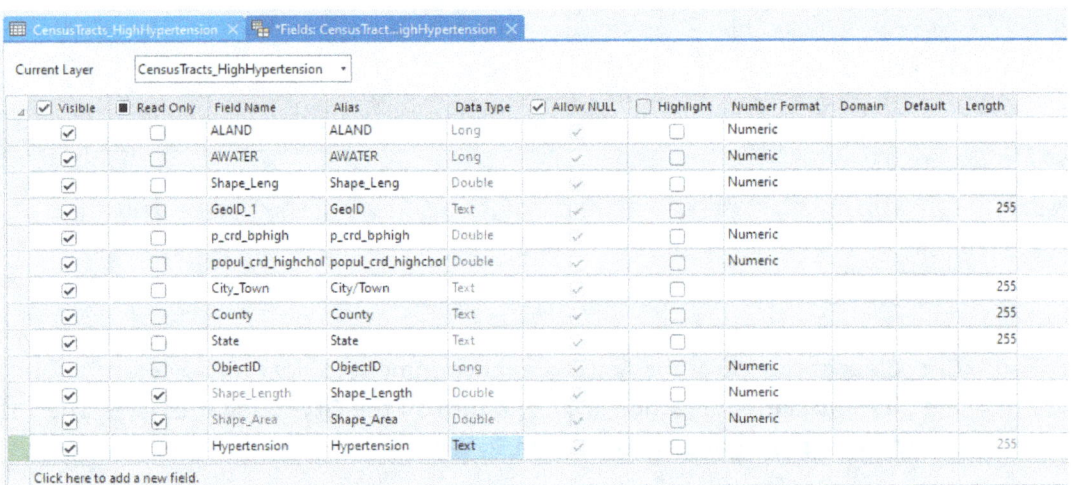

8. Click the **X** in the **Fields:CensusTracts_HighHypertension** tab to close the Fields View. A popup window will appear asking if you would like to save changes. Click **Save**. If you scroll all the way to the left of the **CensusTracts_HighHypertension** attribute table, you will see a new column labeled **Hypertension** that has all null values. Here we will use **Select By Attributes** to fill in these cells. First, you will click **Select By Attributes**, and choose where **p_crd_bphigh is null**. Hit **Apply** then **OK**.

104

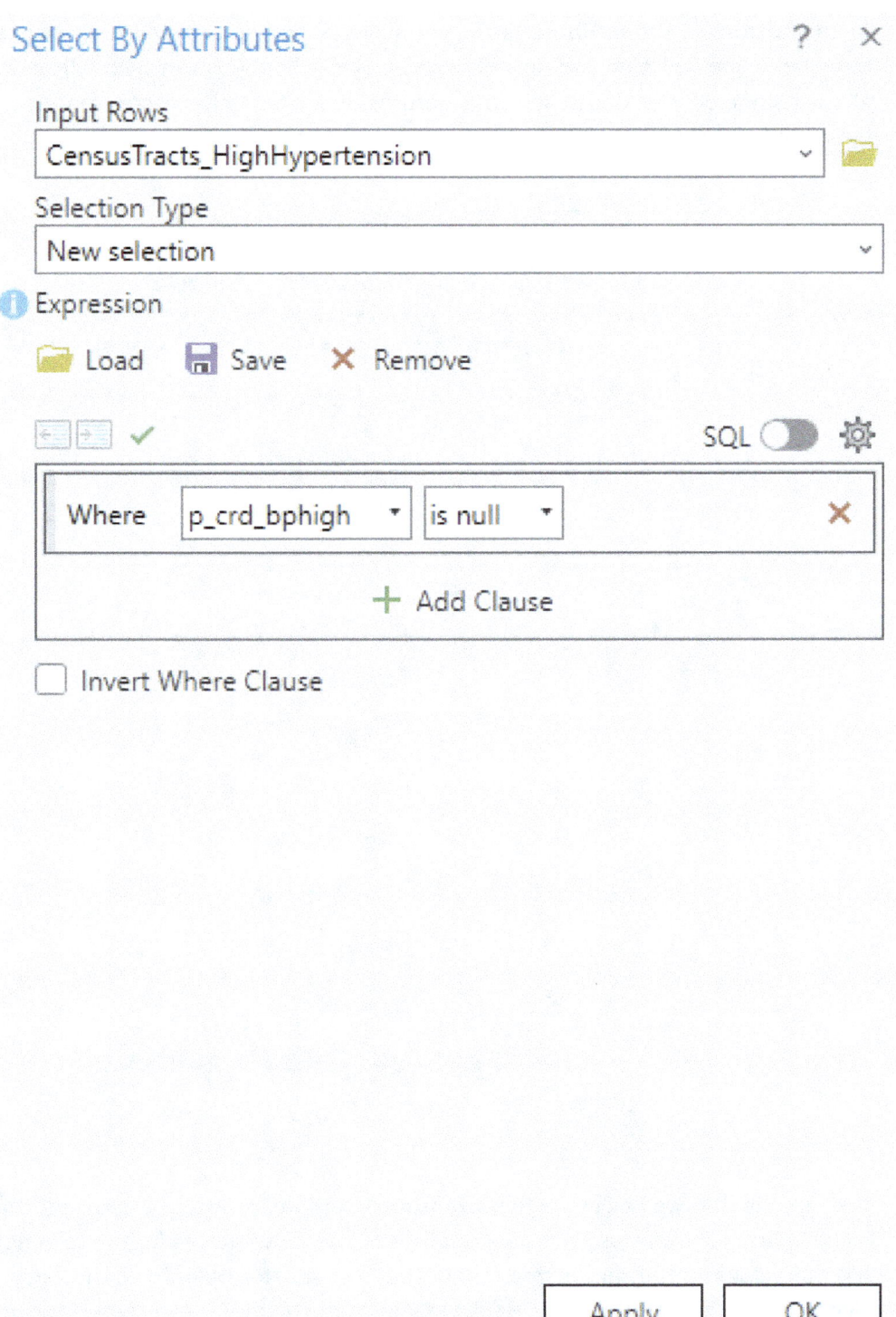

9. At the bottom of the attribute table, you will see figures showing rows that alternate between white and blue and only blue rows. This is how you can switch from looking at all the data to only looking at selected features. To look at only selected features, click the figure with only blue rows.

	OBJECTID_1 *	Shape *	STATEFP	COUN
1	1	Polygon	47	157
2	2	Polygon	47	157
3	3	Polygon	47	157
4	4	Polygon	47	157
5	5	Polygon	47	157
6	6	Polygon	47	157
7	7	Polygon	47	157
8	8	Polygon	47	157
9	9	Polygon	47	157
10	10	Polygon	47	157
11	11	Polygon	47	157
12	12	Polygon	47	157

219 of 249 selected

10. Because the data we had in the **Hypertension** excel spreadsheet only gave information for high hypertension census tracts, we can take all these null values for the selection (which has null values for the census tracts) and label them as **Not High**. To do this, right-click the **Hypertension** field and choose **Calculate Field**. In the textbox under **Hypertension =** enter "**Not High**". Then click **Apply**.

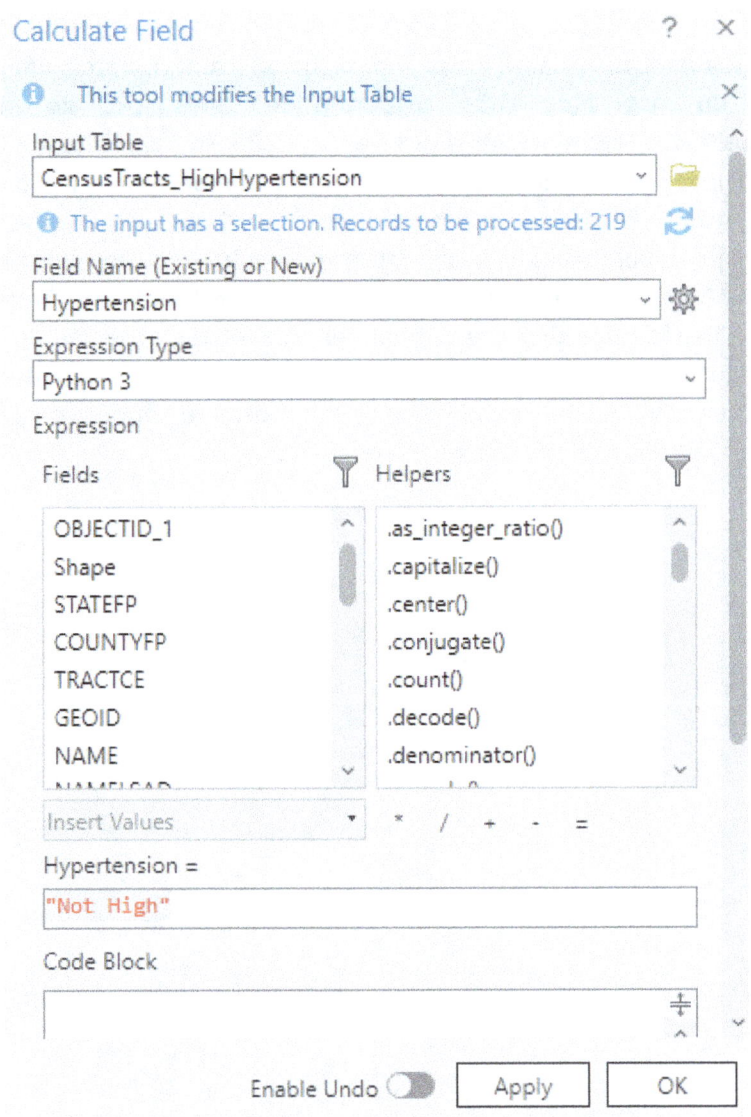

11. Next, you will click **Switch** at the top of the attributes table to switch the selection. This will make the selection change from all null values in the **p_crd_bphigh** column to the selected features whose **p_crd_bphigh** column is not null. You will notice that the **Hypertension** field for the newly selected features all have null values. We can assign them **High** by right-clicking the **Hypertension** field and choose **Calculate Field**. In the textbox

under **Hypertension** = enter **"High"**. Then click **Apply**. Next click **OK**. Finally, clear the selected features and close the attribute table.

12. Now we want to visualize areas with high hypertension rates different than those areas that do not have high hypertension rates. To do this, right-click the **CensusTracts_HighHypertension** layer and choose **Symbology**. Choose the dropdown menu under **Primary Symbology** and choose **Unique Values**. Next, choose the dropdown menu next to **Field 1** and choose **Hypertension**. Now you have the two classes: High and Not High. Double click the symbol next to **High** use the dropdown menu next to **Color** to choose **Medium Fuschia**. Click the back arrow at the top of the symbology pane to go back. Then, double click the symbol next to **Not High** use the dropdown menu next to **Color** to choose **No Color**. Click the back arrow at the top of the symbology pane to go back to verify it looks like the figure below. Close the symbology pane.

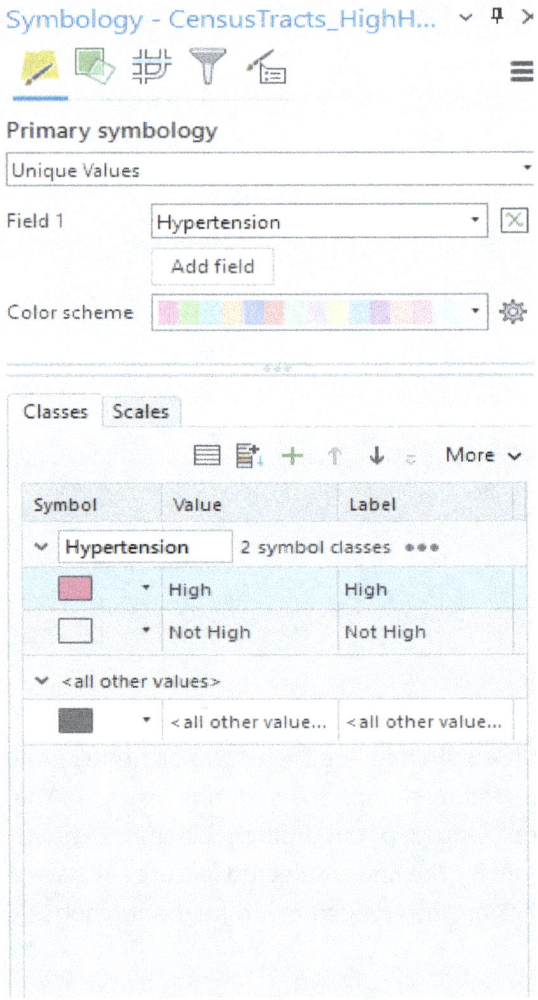

13. Next we will work with labels. We want to label the health clinics a very specific way. First, right-click the **HealthClinics** layer and choose **Labeling Properties**. This brings up a pop-up window on the far right of the ArcGIS Pro environment. This process can be a bit more involved than other because you have to create an **Expression**. You can do this by double-clicking the field you want to use for the label, but you have to make sure that you delete all text in the **Expression** first. Let's try it. Delete all the text in the **Expression** textbox, then double-click the **Practice/Partner Organization** under the **Fields box**. Click **Apply**.

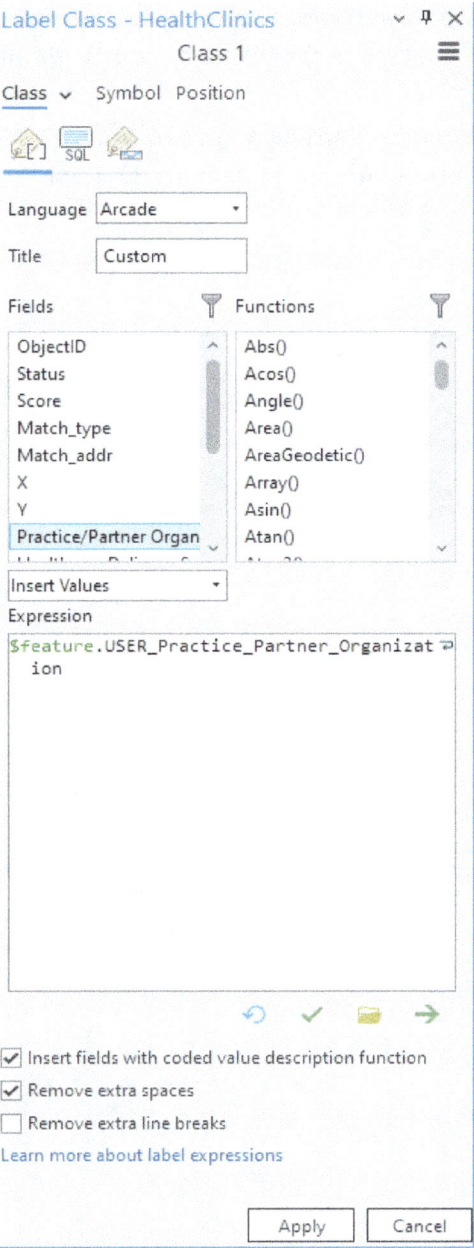

109

14. As you can see, the code in the **Expression** box now says "$feature.USER_Practice_Partner_Organization". If you right-click the **HealthClinics** layer, you will now see the name of the organization by each point location. This is a good start to labeling the features. However, viewers might want to know not only the organization's name, but the health delivery system for that organization. To add this in and make it readable for the viewers, you would want to put the health delivery system in parentheses after the organization. This requires a more involved expression as there is no easy button click to include make parentheses around text. However, you can do it by adding a bit of your own coding to the **Expression** box. Delete everything inside the **Expression** textbox and type in the following code without leaving any spaces:

$feature.USER_Practice_Partner_Organization + ' (' +
$feature.USER_Abbreviations_of_Health_Deliver + ')'

Then, click **Apply**.

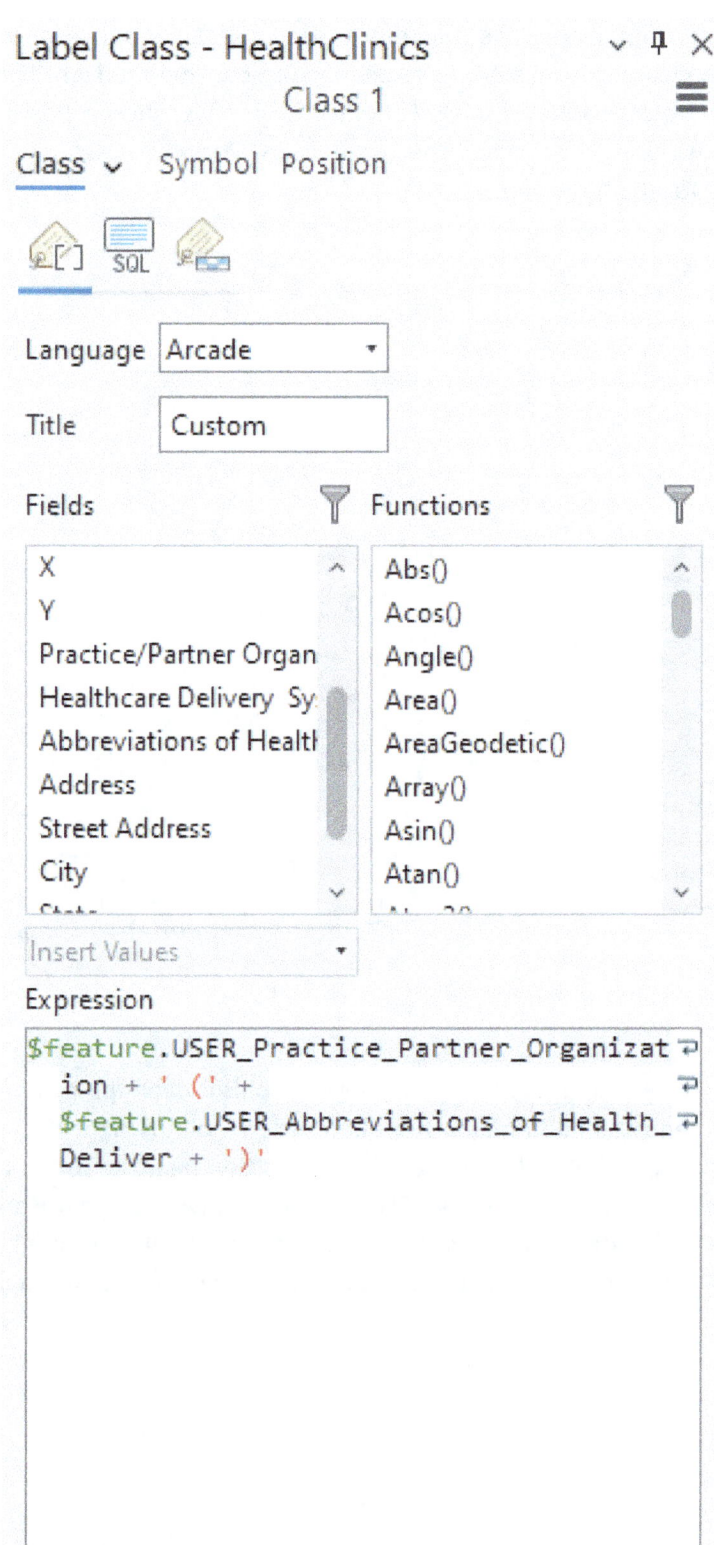

15. This is starting to make the map much more appealing but it still looks kind of busy. Let's finish by cleaning it up a bit. First, change the basemap to **Light Gray Canvas**. Then right-click **CityBoundaries** and select **Labeling Properties**. Choose **Position** from the top of the pop-up window. Next, select the box next to **World Light Gray Reference** to turn off the labeling for the basemap.

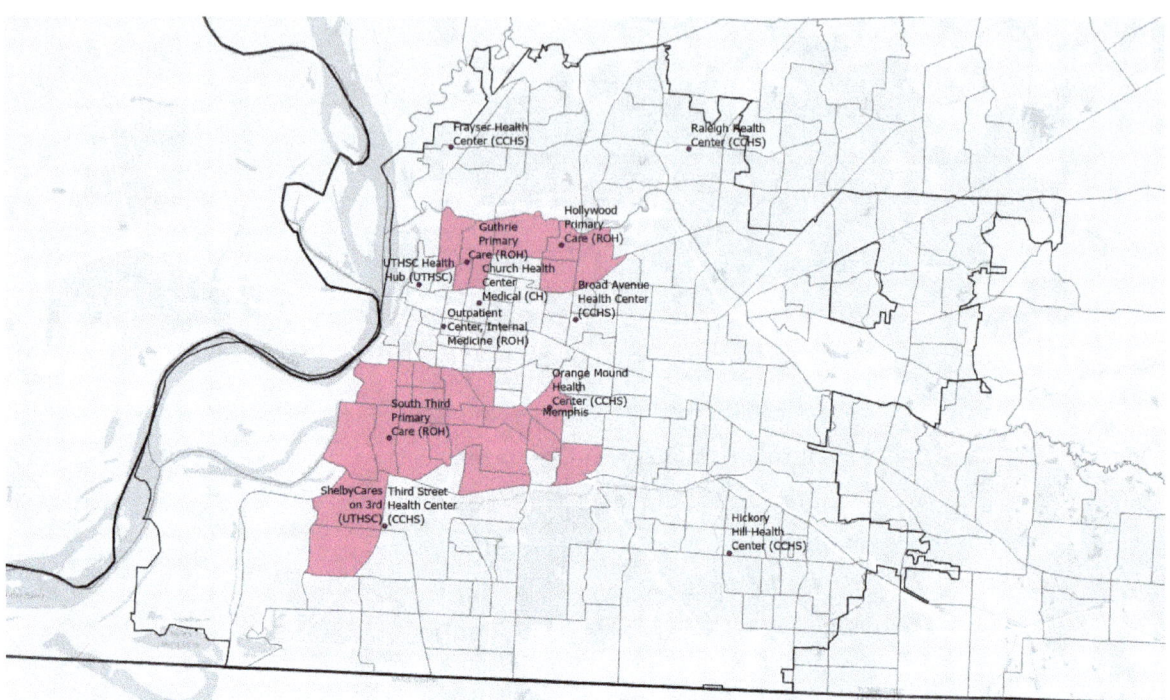

16. This looks cleaned up, but the label for Memphis is overlapping one of the health centers. We can reposition and make this label more visible by right clicking the **Memphis** layer and choosing **Labeling Properties**. Select **Position** near the top of the pop-up window for the labeling properties window. Click the drop-down menu for **Horizontal in polygon** and choose **Horizontal around polygon**. Look at the wheel with numbers. This allows you to choose where you would prefer the label to appear. Change the values to reflect the figure below.

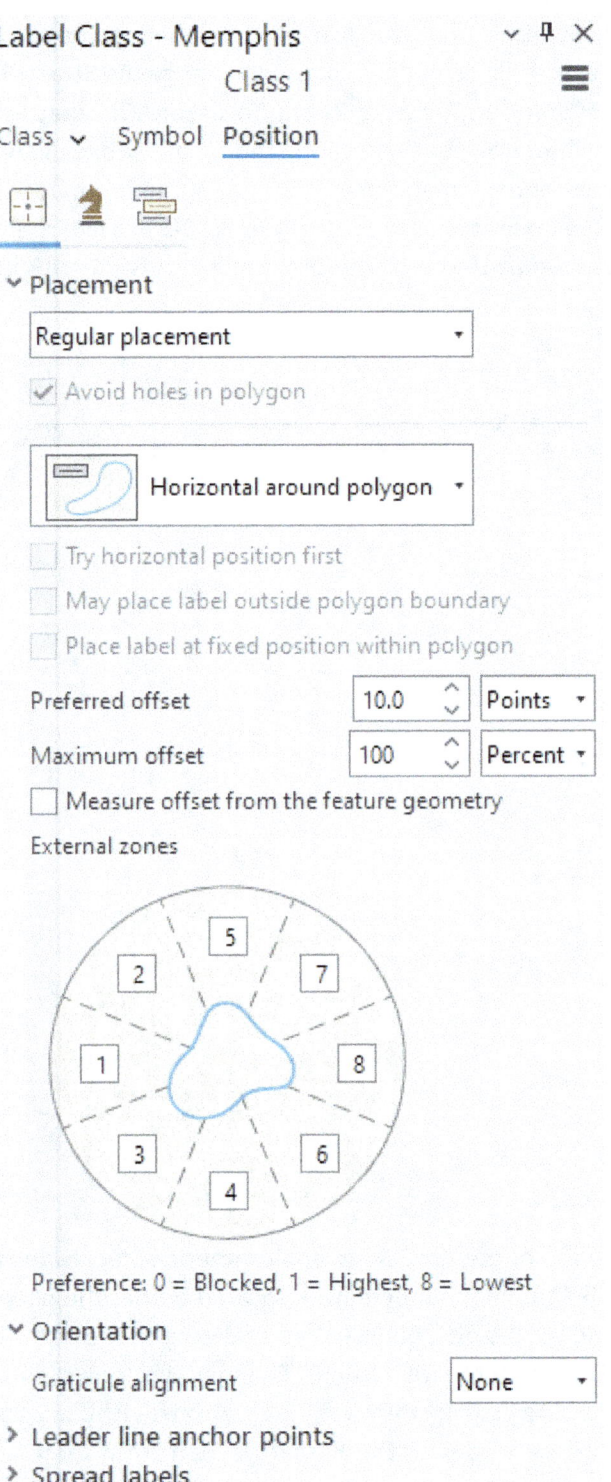

17. Then, click **Symbol** near the top of the labeling properties pop-up window. Under **Appearance**, use the drop-down menu to change the **Size** to **20 pt**. Click **Apply**. Next right-click **HealthClinics** and choose **Zoom To Layer**. The map is almost finished, all that is needed is a **Title**. However, you cannot add a title while in an active **Map** frame. You must create a layout for the map. ArcGIS Pro has some pre-made layouts you can choose from by clicking the **Insert** tab at the top of the ArcGIS Pro environment and selecting the arrow below **New Layout**. Under **ISO – Landscape** choose the **A3** layout. A new blank page will appear.

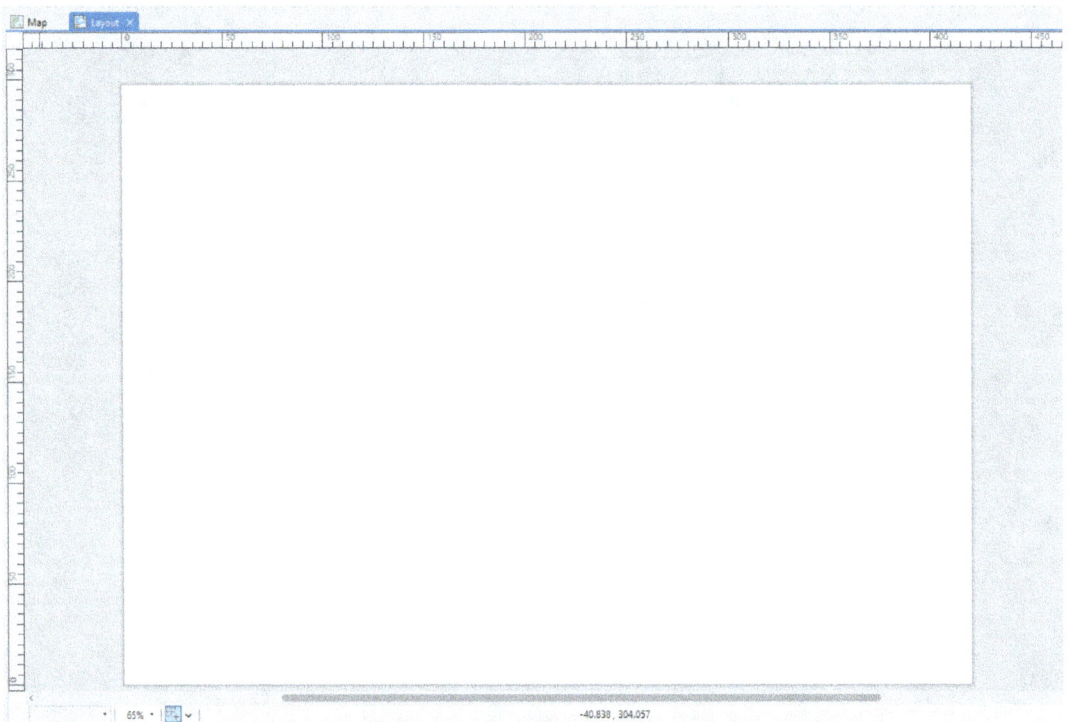

18. Towards the top of the ArcGIS Pro environment, select **Map Frame** and choose the image that looks like the map you created (located within the **Map** section). Then, click at the top left corner of the new blank layout and drag to the bottom right corner to encompass the entire page with the map frame. The map will look slightly off from what you created in the **Map** tab. You can fix this by right-clicking **HealthClinics** and choosing **Zoom To Layer**. Next, you will want to click the **Rectangle Text** from the **Graphics and Text** menu along the top of the ArcGIS Pro environment.

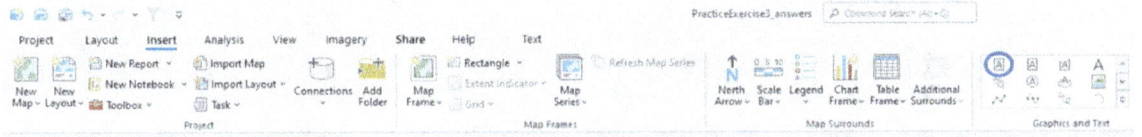

19. Next, click at the top-left of the map layout and drag down and to the right until right before the point location for **Frayser Health Center (CCHS)**. Then double-click in the text box and an **Element** pop-up window will appear on the right-hand side of the ArcGIS Pro environment. Within this pop-up window, click inside the text box and delete **Text**. Then type in **Census tracts with crude adult hypertension prevalence of 53% or higher**. Change the **Margin** to 2 mm. Click the box with a paintbrush (when you hover over it, it will say **Display**) Click on **Text Symbol** towards the top of the pop-up window. In the **Background** section, click the drop-down menu across from **Symbol** and choose **Arctic White**. Click the **Text Symbol** tab and for **Size, choose 18 pt**.

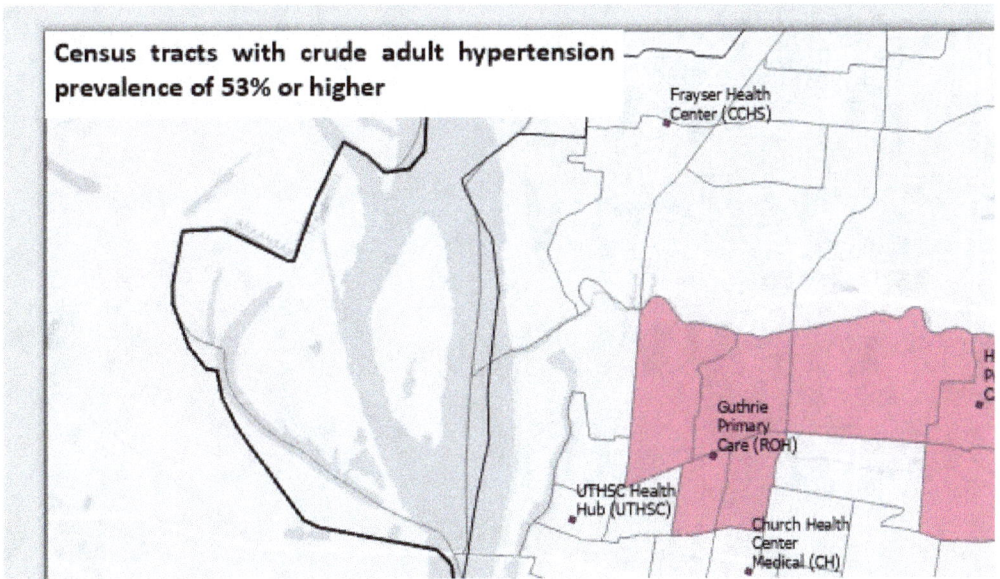

20. On the map layout, resize the text box, so everything fits on one line. Move the text box, so it is not touching the border of the map layout.

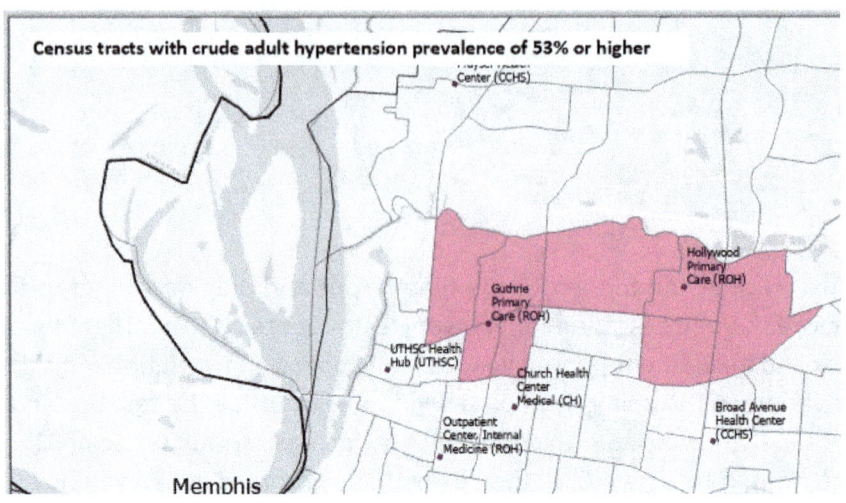

As you can see, the title is now covering over the name of **Frayser Health Center**. You can zoom out a little on your map to make sure that all health centers and census tracts are visible. At the bottom left of the map, there is a ratio number in a text box. Change that to saying **1:78,000**.

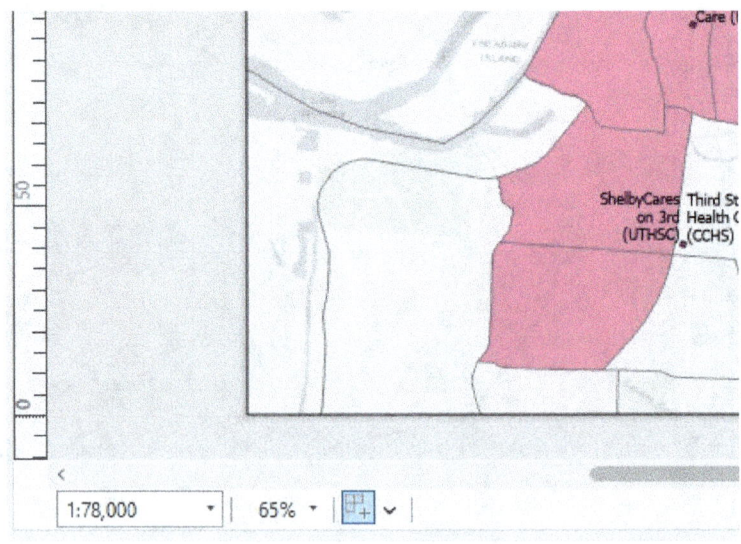

21. Next you will want to save your map as an image on your computer. Click the **Share** tab at the top of the ArcGIS Pro environment, then choose **Export Layout**. In the pop-up window on the right-hand side, make sure that **File Type** is **PNG**. Click the folder button next to the name textbox and navigate to the folder you are saving your exercise answers in. Name it **Final Hypertension Map**. Then click **Export**.

Section 3.1 Task: Submit the **Final Hypertension Map.png** along with your exercise answers.

CHAPTER 4: BIRTH HEALTH

Mastering the Concepts

This chapter explores the use of GIS in analyzing birth health outcomes, focusing on data sets related to birth morbidity, such as low birth weight, birth defects, and infant mortality. It highlights challenges in integrating this data into GIS and examines spatial patterns of birth health. The chapter also delves into the underlying factors influencing these outcomes, including environmental risk factors and healthcare access, and discusses how GIS can help uncover these spatial relationships to improve public health interventions.

Geographic information systems (GIS) are valuable tools for examining the influence of neighborhoods on adverse birth outcomes, such as low birth weight and birth defects. These neighborhoods often share common environmental factors that can affect health outcomes, extending beyond individual risk factors. GIS is increasingly used to monitor how the maternal environment impacts birth health. Spatial analysis helps identify challenges such as limited access to healthcare and potential environmental or behavioral risk factors. Techniques like cluster analysis, probability mapping, spatial filtering, and Bayesian smoothing are used to map variations in birth health, address data instability, and ensure confidentiality. Additionally, GIS helps assess the role of neighborhood risk factors in shaping birth outcomes.

Birth morbidity data is typically collected from electronic birth certificates and vital records maintained by state health departments. Some regions also use systems to gather information on women's experiences before, during, and after pregnancy. Many countries have established surveillance systems to monitor birth health and related conditions. In some areas, birth defects monitoring is also used to assess health risks, particularly in relation to environmental factors like waste management. Hospital discharge data, often collected at the zip code level, are also used in research on birth outcomes, though they may have limitations for population-based studies. Exposure studies commonly rely on birth records, birth defects registries, disease data, or local clinics.

Most countries conduct national surveys that include maternal health questions alongside assessments of nutrition, chronic conditions, injuries, and healthcare use, such as the U.S. National Health and Nutrition Examination Survey and National Health Interview Survey. Unlike standard vital statistics, local surveys focus on broader maternal health issues, including mental health, stress, and family or neighborhood characteristics, though respondents' addresses are not typically released.

In the U.S., birth records provide maternal and infant health indicators, such as birth weight, gestational age, congenital malformations, and maternal factors like age, race, education, prenatal visits, and substance use. A limitation of these records is the quality of maternal data, with inconsistencies in recalling events like tobacco or alcohol use. More detailed data on smoking, drinking, diet, family history, and environmental exposures would improve data quality.

Mother's residential address serves as the geographic identifier on birth records. GIS geocoding converts tabular data into geographic coordinates by matching addresses to a digital street file. Inaccurate or incomplete address information, such as P.O. Boxes or rural routes, reduces matching accuracy and requires manual correction.

When birth records are matched with addresses, census data such as tracts, block groups, and blocks are linked to the records. This enables analysis of birth outcomes in relation to environmental and socioeconomic factors. GIS helps uncover potential links between environmental exposures and adverse birth outcomes. However, the mother's current address may not reflect the environment during pregnancy, as mothers live in overlapping environments like home, work, or school. These exposures can vary over time, and lifelong environmental factors may contribute to birth defects or infant mortality.

GIS plays a pivotal role in analyzing birth health outcomes, including low birth weight, infant mortality and birth defects, by offering spatial insights that link demographic, environmental, and healthcare factors. Low birth weight (LBW), defined as infants weighing less than 2500g, is a major concern due to its link to higher mortality and long-term health risks. GIS plays a critical role in studying LBW by mapping maternal residential addresses alongside demographic and socioeconomic data. It enables the analysis of birth weight variations within small geographic areas that may share environmental and social characteristics, such as pollution and healthcare access. GIS tools help researchers adjust for factors like maternal age and assess LBW rates in areas with similar environmental exposures.

Infant mortality data, derived from death certificates, provide vital information on demographics, causes of death, and contributing factors. Death certificates contain two types of address data: the place of death (e.g., hospital or nursing home) and the decedent's residence. The residence address is particularly useful for understanding environmental influences on health, while the place of death can highlight healthcare disparities, particularly within different racial and ethnic communities.

Neighborhood context is a crucial risk factor for infant mortality, with disadvantaged areas often experiencing higher rates. These neighborhoods typically face common challenges such as unemployment, substandard housing, crime, inadequate healthcare, and environmental pollution. Infants born in these areas are at increased risk due to factors like prematurity and low birth weight. Racial disparities in infant mortality, as well as varying birth outcomes among African immigrant and Mexican American populations, suggest that social and environmental factors—rather than genetic predispositions—play a significant role in shaping these outcomes.

GIS tools are invaluable in examining these complex relationships. GIS can help map and identify "hot spots" where poverty, violence, and environmental hazards intersect. However, comprehensive GIS analyses that account for the interconnectedness of demographic, socioeconomic, and environmental factors remain rare. Aggregating individual birth and death records by geographic units, such as census tracts, allows for a clearer understanding of spatial patterns and risk factors.

To deepen the understanding of these spatial dynamics, GIS is also used to compute continuous spatial distributions of infant mortality rates. One method, "punctual kriging," involves defining a region around each location and calculating the infant mortality rate within it. By repeating this process across a grid of locations, GIS generates isarithmic maps that highlight areas with significantly high mortality rates, offering a visual representation of risk zones.

Despite challenges, GIS-based studies have proven effective in assessing environmental exposures linked to birth defects. By creating composite databases of contaminant sources and evaluating how contaminants disperse through air or groundwater, GIS helps identify areas at risk. Chemical dispersion models, integrated with GIS, visualize the spread of contaminants across different spatial scales. These models overlay contamination footprints with census and birth morbidity data to map populations at risk, helping evaluate exposure sensitivity.

In addition to environmental factors, GIS analysis enhances understanding of maternal health at various geographic scales, from the individual to the global level. It supports better prevention, intervention, and access to health services. However, effective GIS applications depend on the availability of accurate birth outcome data, such as birth defect registries and vital records. Integrating geographic data into health surveys can further enhance the utility of GIS. Despite these advances, challenges persist in collecting and integrating maternal health data across different administrative units and time periods, particularly when linking birth events to environmental contaminants.

Mastering the Skills

Exercise 4: Birth Health Outcomes – Creating a Health Story using ArcGIS StoryMaps

In this tutorial you are going to become familiar with using health data to create a story using ArcGIS StoryMaps. This will provide you with experience in creating interesting and easily understandable health data visualizations that can be readily sharable to others, including the general public.

OBJECTIVES

- Use health data to create an interesting story
- Make the information in the story easily understandable
- Share the story with the general public

Required data:

All data, preconfigured GIS project files, and geodatabases required for this exercise are already included with the tutorial data and are available through the Resources page of the author's official website: www.esraozdenerl/resources. Instructions for accessing the tutorial data can be found in the Tutorial Data section of the Preface.

Data sources:

1. Data from the Birth Health Outcomes folder. Sources:
    a. Pregnancy resource centers (Shapefile created from geocoded addresses found through a web search): Since locations may change, you can update the addresses in the provided Excel file using a simple Google search.
    b. Informational resources on low birthweight

2. You will also need:
    a. Outside resources you research yourself

Section 1: Creating a Health Story using ArcGIS StoryMaps

Section 1.1: Use health data to create an interesting story

There is a lot of health information out there. When presented in an interesting and easily understandable way, the general public will be able to better understand what is being portrayed and how they can help themselves and others experiencing certain health issues.

For the first part of this tutorial, you will learn how to make an interesting and easily understood story map to talk about birth health outcomes. This exercise is meant to be vague to allow your creative thinking to bring on a unique and intriguing story map using ArcGIS StoryMaps. Collect and save eye-catching images to add to your story map and save them in one folder for easy access. These images can be created by you, created by AI software, or collected from the web. Just make sure to cite all sources used.

3. Open ArcGIS StoryMaps home page (https://storymaps.arcgis.com/) and sign in with your ArcGIS Online organizational account.

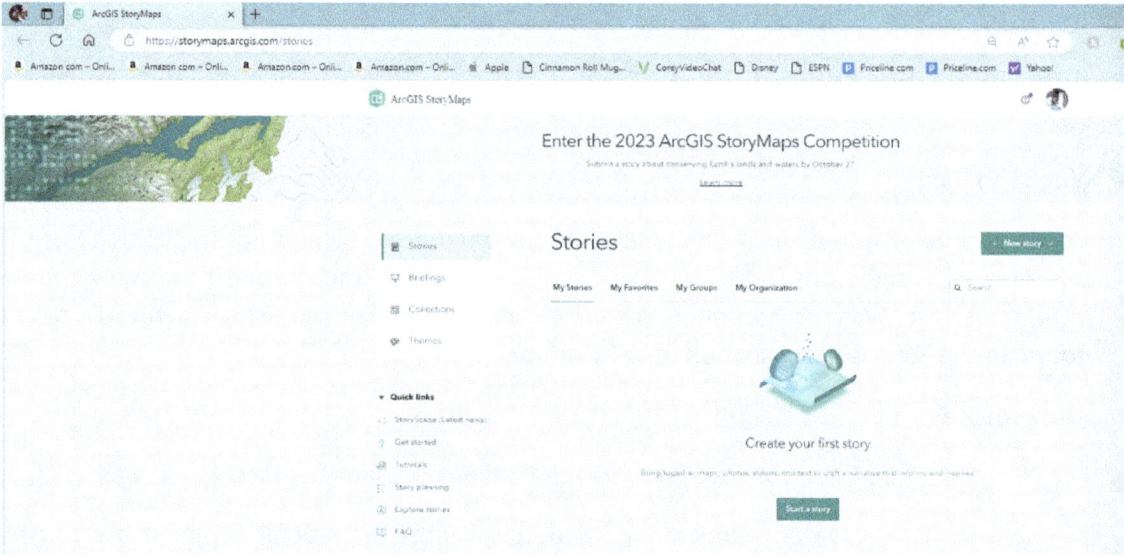

4. Click **New Story** in the upper right of your web browser and choose **Start From Scratch**.

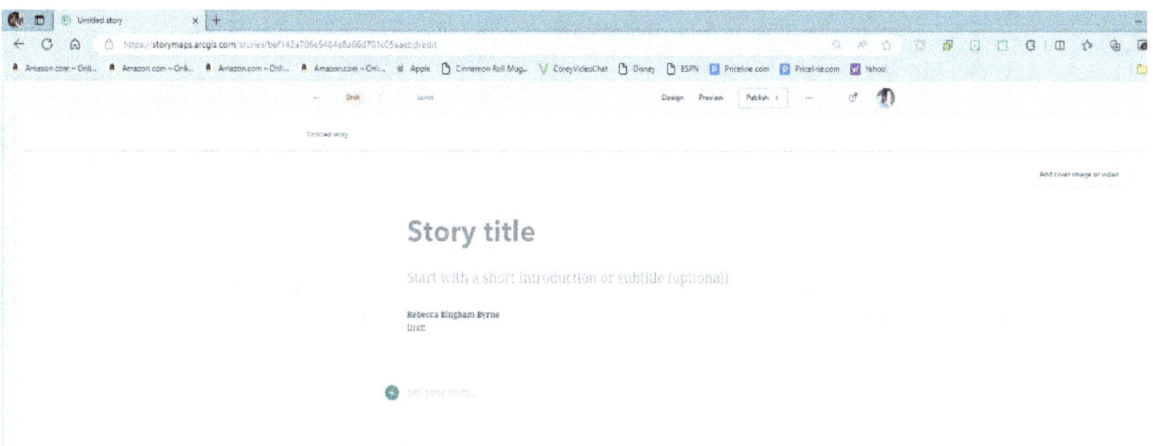

5. Click into **Story Title** and create an interesting and relevant title for your story map. Then click into **Start with a short introduction or subtitle (optional)** to create a subtitle that may be a little more descriptive of your story. After that, click Add cover image or video in the upper right corner, and add the photo or video you decided to use for your cover image.

6. Using the background information provided in the data folder you downloaded, create an **introduction** with background information about low birth weight (i.e., what it is, diagnosis, causes, risk factors, symptoms, treatments, and complications).

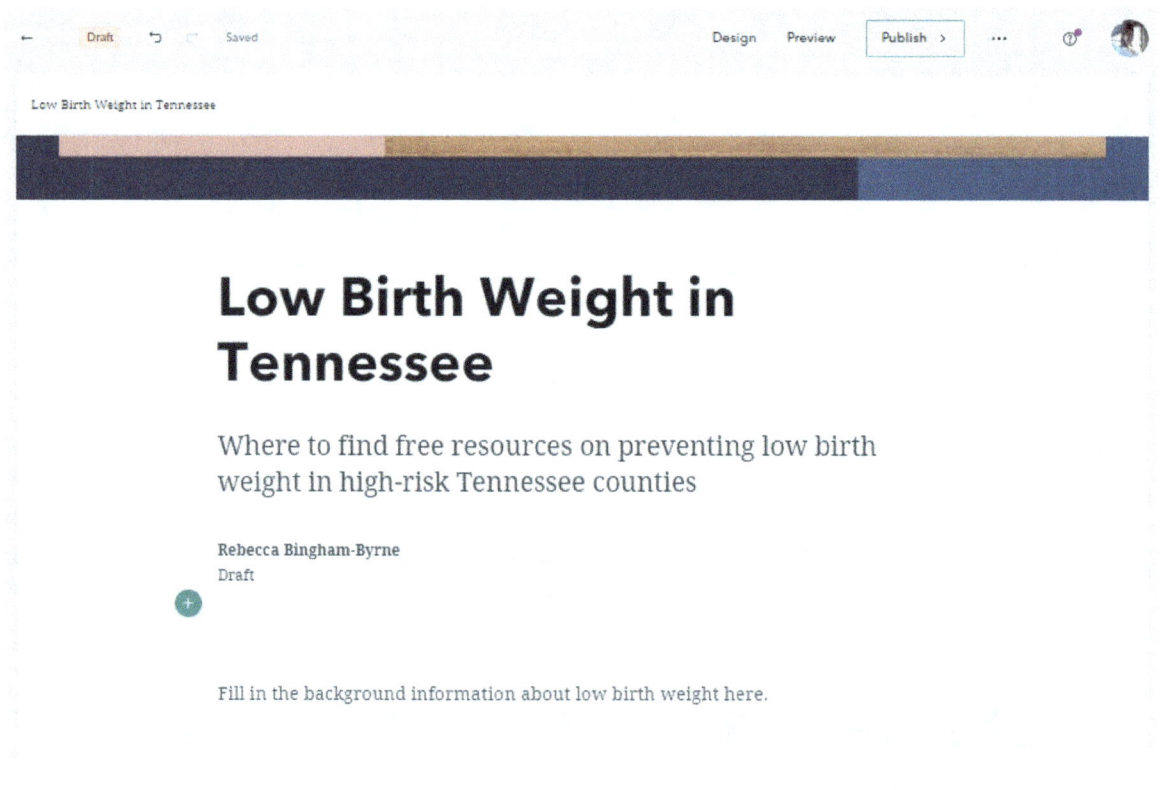

7. Next you want to add the map with areas of interest into your story. You will want to click the **+** button below your introductory text and choose **Map Tour**. Choose **Start From Scratch** then choose **Explorer**. Click **Next**, then choose **List** and click **Done**. This puts in a new blank map of the world.

Low Birth Weight in Tennessee

8. You can edit the basemap by clicking the icon that looks like a pencil on the top center of the map. Choose **Select basemap**, then **Browse more maps**. Click the **My Organization** tab and in the search bar type **Low birth weight (less than 2500g) in Tennessee**. Select this map and change the initial zoom level to **Custom** and use the dropdown menu to choose **States or provinces**. Then choose **Done**.

9. Now you will begin to add tour points. Here, you will use the excel spreadsheet provided in the data folder to add locations for each of the pregnancy resource centers located in counties that have low birth weights for more than 9% of births in the county. Images and resource information you will collect for these pregnancy resource centers on your own to create a unique and interesting map tour for the general public. Click on **Untitled tour point**. Then click on **Add location**. In the search bar, type in the address including city, state and zipcode and hit enter. A location will appear on your map. Click **Add to map**, then choose **Add location**. Now you can edit the **Title** to the name of the pregnancy resource center and add in a description and image. The image can be anything related to the center from the organization's logo to an infographic of their center to a picture of their building. The description should include information about the center such as services they provide or hours of operation. Maybe even a link to their website. Once finished, click the + button at the bottom right of the map tour to add a new location. Edit this data in the same fashion to keep everything consistent.

10. Continue adding locations until each of the 19 pregnancy resource centers have been added and their information has been edited. When finished, click **Publish** in the upper right of the web browser. Set the sharing level to **Everyone (Public)**. Then click **Publish**.

11. **Submit your StoryMap link to earn credit for the exercise.**

CHAPTER 5: INFECTIOUS DISEASES

Mastering the Concepts

This chapter is about GIS and its role in understanding the spatial dynamics of infectious diseases. GIS is a vital tool for investigating how diseases spread through time and space, driven by microorganisms that transmit either directly between individuals or via vectors. The chapter explores the spatial determinants of these pathogens and their interactions, emphasizing that cases of infectious diseases are often geographically dependent. By mapping these patterns, GIS helps track the spread of new and reemerging diseases. It also highlights the importance of global surveillance and the challenges of monitoring infectious diseases, especially in low-resource settings, in order to prevent and control their spread across borders.

Understanding transmission dynamics is key in GIS-based assessments of infectious diseases. Spatial distributions reflect environmental factors that influence risk and susceptibility. For vector-borne diseases, GIS helps analyze the distribution of vectors (often insects), reservoir hosts, and human activity patterns where host-vector-human interactions occur. Non vectored infectious diseases, like HIV/AIDS, tuberculosis, measles, and influenza, spread directly from person to person. These diseases can be airborne, transmitted through respiration, or waterborne.

Waterborne diseases are contracted by drinking contaminated water or through direct contact with contaminated recreational waters. Some diseases, like tetanus, are caused by toxins from food or soil, while sexually transmitted diseases (STDs) like HIV/AIDS and syphilis spread through skin or sexual contact.

A GIS integrates infectious disease models with diffusion models, capturing the space-time dynamics of disease spread. There are four types of disease diffusion patterns: contagious, hierarchical, network, and mixed diffusion. While address data is crucial for GIS-based analysis, data are often aggregated by geographical units (e.g., zip codes) due to privacy concerns. Key data sources for disease reporting include sentinel systems, surveillance systems, and public health surveys. This chapter explores GIS applications in analyzing both vector-borne and non-vectored diseases, highlighting spatial analysis approaches and future implications.

Vector-borne zoonotic diseases require a pathogen, vector, reservoir hosts, and a human victim. Common vectors include mosquitoes and ticks. Examples of mosquito-borne diseases are dengue fever, malaria, West Nile virus, and yellow fever. Tick-borne diseases include Lyme disease, Rocky Mountain spotted fever, Heartland virus, Colorado tick fever, and Southern tick-associated rash illness.

The global spread of vector-borne diseases highlights the need to understand their geographic extent. Over the past decade, GIS and spatial modeling studies have grown, focusing on climate, biological, and ecological factors. With the use of GIS and remote sensing in surveillance, especially for ticks, birds, and mosquitoes, comparability across studies has improved.

For example, most studies use GIS to assess the relationship between Lyme disease (LD) risk variables (e.g., vector, pathogen, host) and environmental factors. Since ticks have limited mobility, host range is a key spatial determinant. When field data is unavailable, tick distribution is predicted based on ecological information. GIS and remote sensing (RS) tools enable predictive modeling of tick distribution using environmental variables like vegetation, climate, and host populations. Some studies have shown GIS-derived data to predict nymphal density better than field data.

Data such as tick densities, pathogen genotypes, and human incidence are mapped using various methods, including density mapping and risk modeling. GIS models incorporate data like soil types, forest cover, and proximity to water to analyze tick habitats. Predictive models use spatial tools to account for climate and ecological variables influencing vector distribution. Models projecting climate change effects on tick populations utilize data from NASA, USGS, CDC, and NOAA.

However, limitations exist in the spatial analysis of LD. Inconsistent case tracking, reporting errors, and boundary issues can skew results. Lack of baseline data on tick surveillance and the high cost of field studies further complicate research. Region-specific approaches are needed to understand LD risk factors, such as sociocultural factors, demographics, and local weather patterns. More standardized data collection and long-term studies are necessary to better understand the spatial variation of LD.

Future research should focus on integrating GIS, RS, and ecological studies to model LD transmission cycles and climate change impacts. Temporal and spatial variability in tick-borne diseases highlight the need for dynamic models that consider local climatic and ecological conditions. Fine-scale human incidence data is crucial for identifying isolated endemic areas. Effective risk mapping and surveillance require robust field studies to validate predictions.

As research advances, GIS and spatial analysis will be key in developing control measures like targeted vaccine distribution and interventions in newly identified endemic areas. Policies could also involve collaboration with the tourism sector to monitor high-risk areas.

Waterborne diseases are caused by pathogenic microbes spread through contaminated water, leading to illnesses like cholera, cryptosporidiosis, and giardiasis. These diseases can spread via drinking water or direct contact with contaminated environmental waters. Water contamination often comes from humans, animals, or industrial sources, posing health risks not only through drinking but also via the food chain.

GIS plays a vital role in mapping global water supply hotspots and analyzing water quality. It uses satellite imagery to identify water features and assess water quality. GIS-based dashboards and surveillance systems track waterborne diseases, providing real-time data to guide public health responses. These systems integrate epidemiological, demographic, and water quality data, offering insights into disease patterns and water supply risks.

GIS techniques, including space-time analysis, help track disease outbreaks, as exemplified by John Snow's 1854 cholera mapping in London. Modern GIS tools, like kernel density estimation and network-based scanning, are used to identify disease clusters and improve surveillance and prevention efforts. GIS also aids in water infrastructure management, helping monitor water availability, distribution systems, and sustainability.

Airborne diseases, including COVID-19, influenza, and tuberculosis, spread through respiratory contact, often exacerbated by poor housing, overcrowding, and close contact in places like schools, shelters, and workplaces. High population density, poverty, and crowding increase transmission risks. Urban areas, for example, have higher rates of tuberculosis and COVID-19 compared to rural regions. GIS-based methods can identify high-risk populations and improve disease control by mapping transmission patterns and strain variations.

During the COVID-19 pandemic, GIS and GIS-based dashboards played a critical role in disseminating real-time data on infection rates, hospitalizations, and deaths. Platforms like the Johns Hopkins University COVID-19 dashboard became globally recognized tools, providing interactive maps and visualizations that helped governments, health organizations, and the public track the spread of the virus. These dashboards utilized GIS to map case densities, monitor hotspots, and highlight trends by region, offering immediate insights that guided policy decisions and resource allocation.

In addition to COVID-19, air pollution contributes to respiratory conditions like asthma, with GIS helping analyze environmental exposures, such as air quality and housing, to understand asthma's impact and its potential interaction with COVID-19. Individuals with pre-existing respiratory conditions may experience worse outcomes from the virus. By integrating various environmental data sources, GIS can identify areas with elevated health risks, such as those near pollution sources or hazardous land uses.

Socioeconomic factors, such as poverty and race, influence asthma and COVID-19 rates, with children and minority or low-income communities particularly vulnerable. However, using aggregated data like hospitalization rates in GIS can limit analysis accuracy, as it doesn't capture individual patient details. Moreover, spatial autocorrelation can complicate traditional statistical analysis. Adjusting for these issues is essential in accurately assessing environmental justice and health disparities in the context of both asthma and COVID-19.

Sexually transmitted diseases (STDs) are spread through sexual contact, and their transmission is difficult to map using traditional spatial patterns. GIS helps examine the spatial distribution of STDs, including clustering and co-infection rates, revealing areas where prevention strategies should be focused. Economic and class disparities affect reporting rates, with higher underreporting in wealthier areas. GIS is used to identify priority populations and locations for STD prevention and intervention, particularly for HIV/AIDS. It also aids in understanding HIV transmission drivers and improving healthcare accessibility in high-burden areas.

Drug addiction and mental health behaviors are closely intertwined, and GIS analysis and spatial models are key tools in studying their epidemiology, linking social interactions, physical distance, and behavioral patterns. GIS is used to assess factors like proximity to treatment centers, risky areas, and drug markets, as well as mental health service facilities. By calculating distances and creating service area buffers, GIS helps understand the relationship between drug users, those with mental health conditions, and their environments. GPS devices collect data on locations like treatment centers and users' residences, allowing for analysis of their proximity to services. GIS supports linking neighborhood data with individual behaviors, enabling better understanding of how neighborhood characteristics influence both drug use and mental health patterns.

Mastering the Skills

Exercise 5: Spatial statistics of Health Data in ArcGIS Pro – Clustering of values to search for spatial patterns

In this tutorial you are going to become familiar with clustering techniques to understand spatial patterns of infectious disease.

OBJECTIVES

- Utilize exploratory data analysis techniques to search, characterize, and describe the spatial distribution of infectious disease incidence data
- Utilize widely used methods to characterize spatial associations at global and local levels (Getis-Ord G Statistics, Global Moran's I, Getis-Ord G*, Anselin Local Moran's I)

Required data:

The original data sources listed below are provided for reference only. You do not need to download or curate these datasets from the original sources. All data, preconfigured GIS project files, and geodatabases required for this exercise are already included with the tutorial data and are available through the Resources page of the author's official website: www.esraozdenerl/resources. Instructions for accessing the tutorial data can be found in the Tutorial Data section of the Preface.

Original data sources:

1. Data from the Practice Exercises 4 folder. Sources:
    a. Lyme disease incidence data was derived from the following websites:
        i. Lyme disease case data between 2000-2017 (https://www.cdc.gov/lyme/stats/survfaq.html)
        ii. Population estimates (https://www.census.gov/programs-surveys/popest/data/data-sets.All.html)
            1. Population Estimates of Resident Population for Counties and States: April 1, 2000 to July 1, 2010
            2. County Population Totals and Components of Change

Section 1: Global Level

Understanding how disease spreads is an important topic for health geospatial researchers. Tobler's first law of geography is "everything is related to everything else, but near things are more related than distant things". This lab begins searching for spatial autocorrelation of data. Strong spatial autocorrelation means that adjacent geographic objects are strongly related to one another (positively or negatively).

Section 1.1: Calculate spatial autocorrelation using Getis-Ord General G

For the first part of this tutorial, you will calculate the Getis-Ord General G to determine if the data is clustered and the cluster's pattern type.

1. Open the **Lyme_disease.aprx** map file by navigating to the Practice Exercises 5 folder and double-clicking the file.

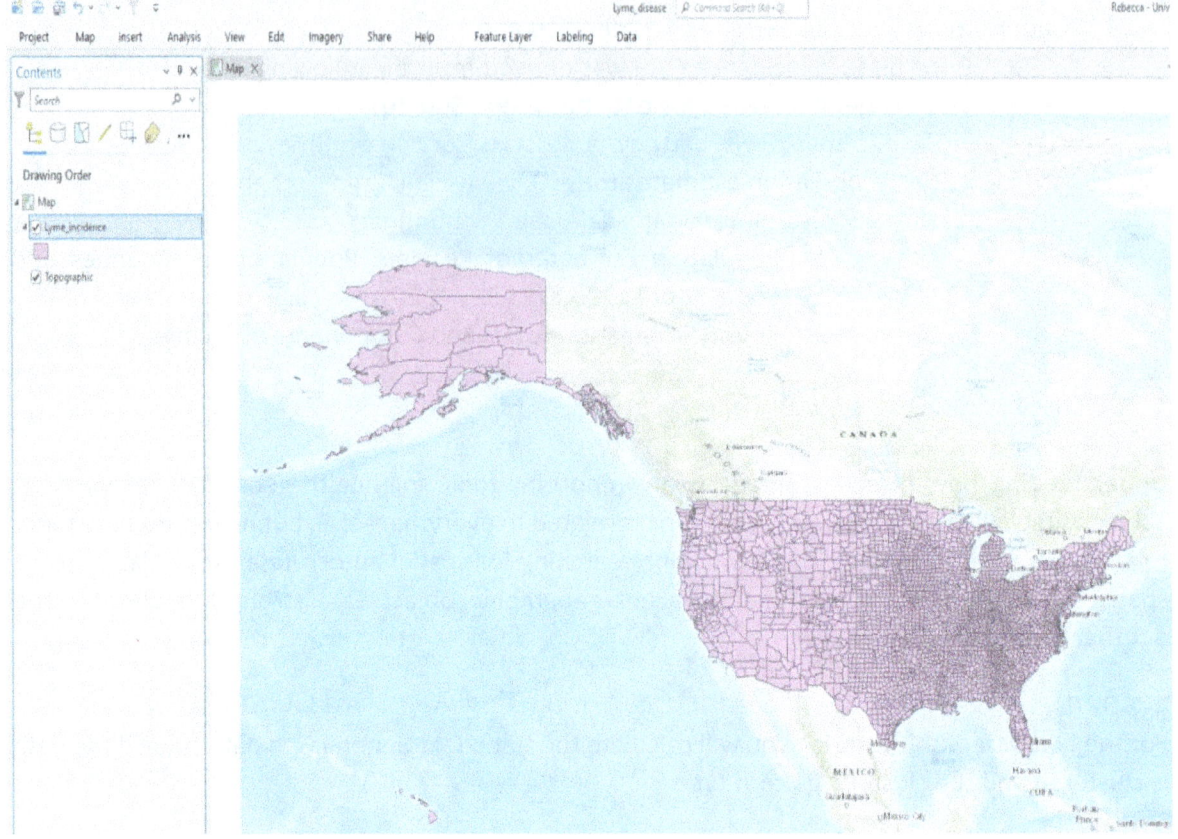

2. Next, you will have to set the data source for each layer in the map file as you did in Exercise 2. A short example of steps is included below. Do this for each layer.
    a. Right-click the **Lyme_incidence** layer->Choose Properties->Select Data->Click Set Data Source
    b. Navigate to the **Lyme_disease.gdb** geodatabase in the data folder for this exercise
    c. Choose the **Lyme_incidence** layer
    d. Click **OK**
3. Next, you will want to click **Analysis** in the top panel of the ArcGIS Pro environment. Click **Tools** to open the Geoprocessing toolbox on the left side of your ArcGIS Pro environment. Navigate to **Spatial Statistics Tools ->Analyzing Patterns->High/Low Clustering (Getis-Ord General G).**

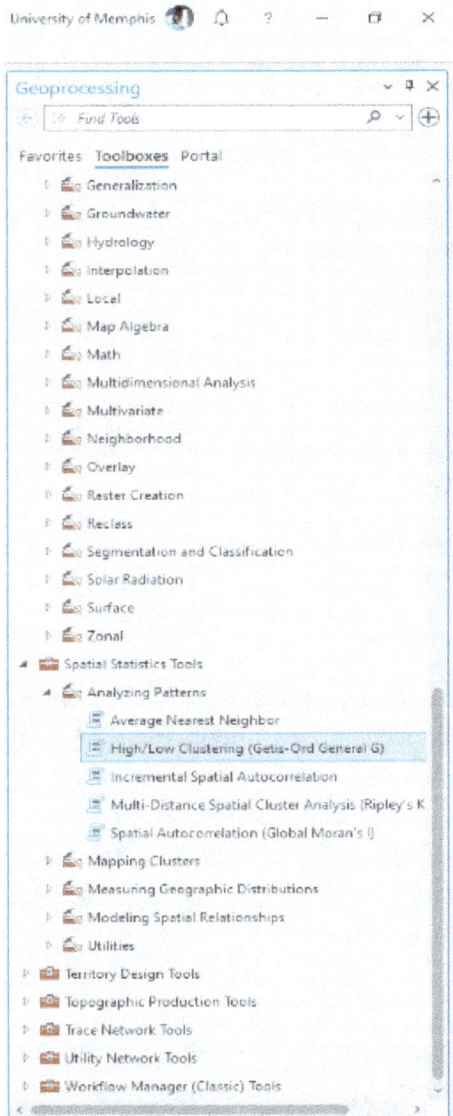

4. Set the input feature class to **Lyme_incidence** and the input field to **AverageInc** (Average incidence rates for the years 2000-2017). Confirm that the conceptualization of distance is set to **Contiguity edges only**. Check the box next to **Generate Report**. This will add graphical outputs to your results window HTML format. Leave all other fields blank and click **Run**.

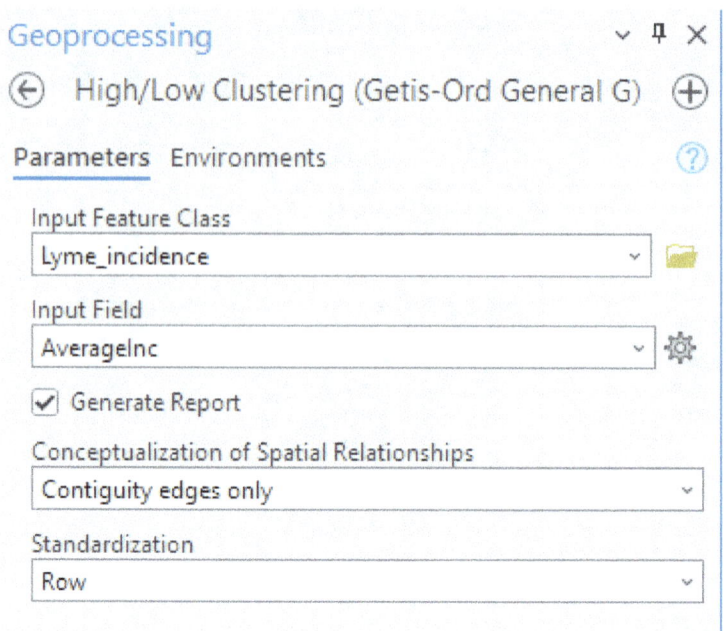

5. Repeat step 4 with input as **Inciden_15** (Lyme disease incidence in 2015), **Inciden_16** (Lyme disease incidence in 2016), and **Inciden_17** (Lyme disease incidence in 2017).

Section 1.1 Task: Create a table by carrying over the output values, including columns for Observed General G, Expected General G, Variance, Z-score, p-value, and Pattern Type for each input (Average, 2015, 2016, and 2017).

Section 1.2: Calculate spatial autocorrelation using Moran's I

For the first part of this tutorial, you will calculate Moran's I to determine if the data is clustered and the cluster's pattern type.

1. In the Geoprocessing toolbox on the left side of your ArcGIS Pro environment, navigate to **Spatial Statistics Tools ->Analyzing Patterns->Spatial Autocorrelation (Global Moran's I)**.

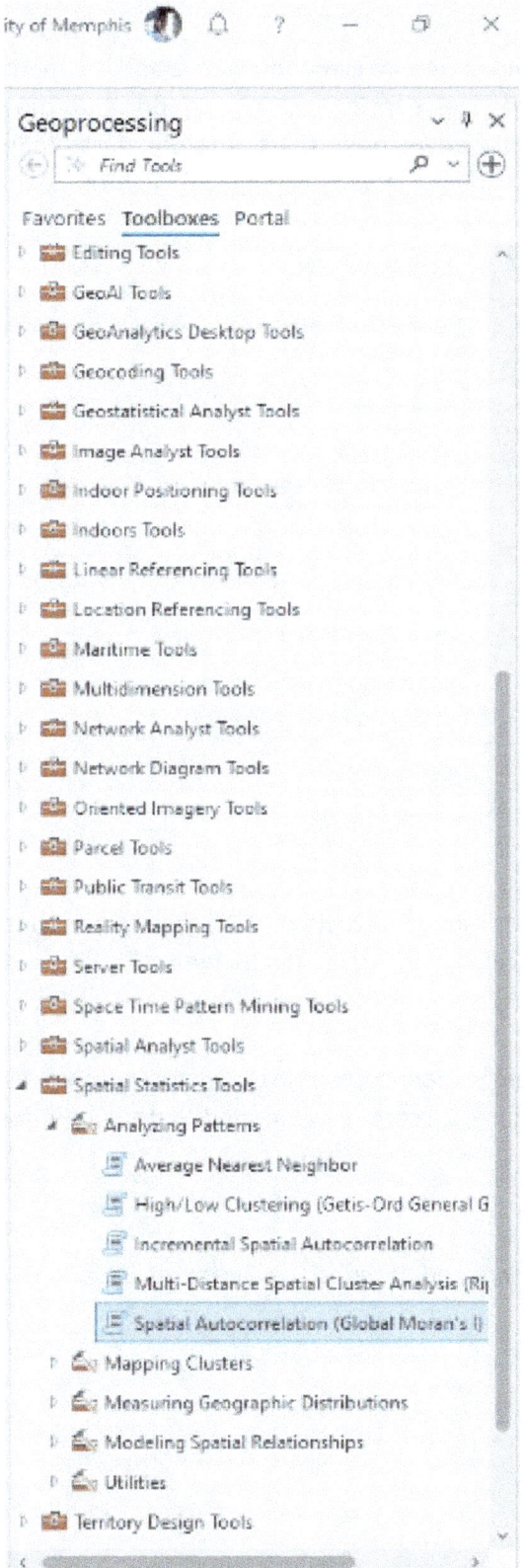

2. Set the input feature class to **Lyme_incidence** and the input field to **AverageInc** (Average incidence rates for the years 2000-2017). Confirm that the conceptualization of distance is set to **Contiguity edges only**. Check the box next to **Generate Report**. This will add graphical outputs to your results window HTML format. Leave all other fields blank and click **Run**.

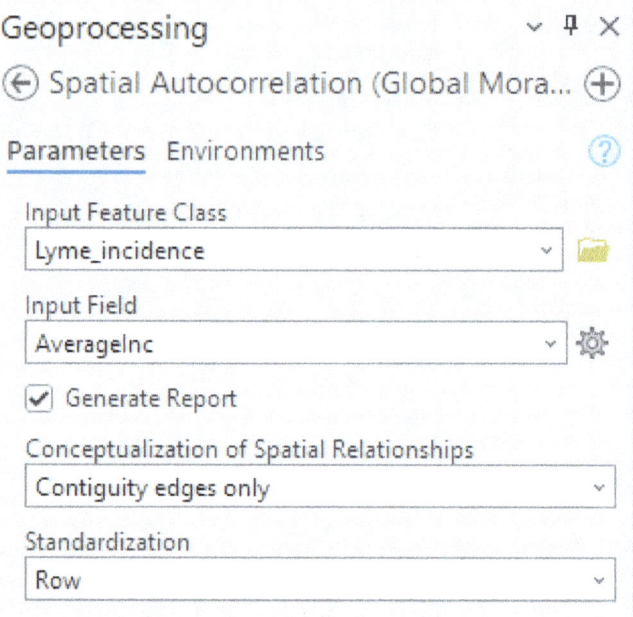

3. Repeat step 2 with input as **Inciden_15** (Lyme disease incidence in 2015), **Inciden_16** (Lyme disease incidence in 2016), and **Inciden_17** (Lyme disease incidence in 2017).

Section 1.2 Task 1: Create a table by carrying over the output values for the Moran's I (Index), Expected I (Index), Variance, Z-score, p-value, and Pattern Type for each input (Average, 2015, 2016, and 2017).

Section 1.2 Task 2: Compare and describe the Getis-Ord General G and Moran's I results.

Section 2: Local Level

Section 2.1: Calculate spatial autocorrelation using Anselin Local Moran's I

For the first part of this section, you will calculate Anselin Local Moran's I to determine if the data is clustered and the cluster's pattern type.

1. In the Geoprocessing toolbox on the left side of your ArcGIS Pro environment, navigate to **Spatial Statistics Tools ->Mapping Clusters->Cluster and Outlier Analysis (Anselin Local Moran's I).**

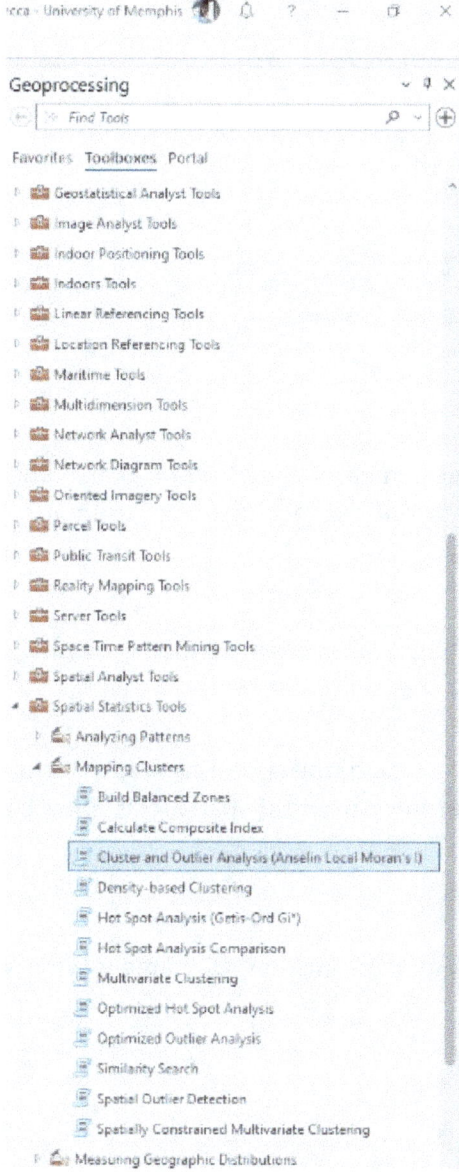

2. Set the input feature class to **Lyme_incidence** and the input field to **AverageInc** (Average incidence rates for the years 2000-2017). Click the folder next to the Output Feature Class field input box. Navigate to the **Lyme_disease.gdb** and name the output feature class **Lyme_incidence_LocalAverage.** Confirm that the conceptualization of distance is set to **Contiguity edges only**. Leave all other fields blank and click **Run**.

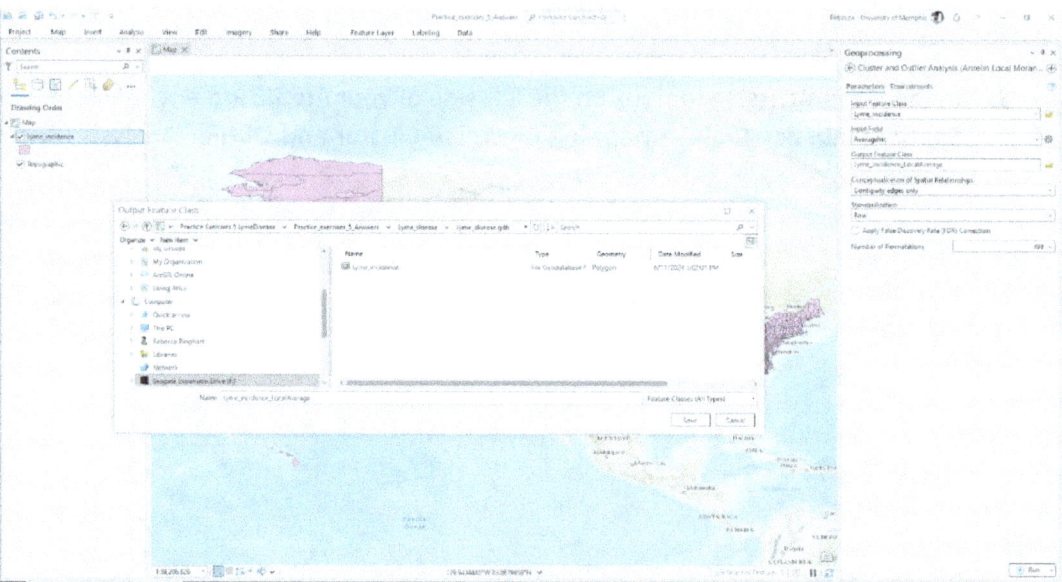

Section 2.1 Task 1: Provide a map of the clustering results for the average incidence of Lyme disease. Describe the pattern of spatial clustering including which areas exhibit clustering and which do not.

3. Repeat step 2 with input as **Inciden_15** (name it **Lyme_incidence_Local2015**).

Section 2.1 Task 2: Provide a map of the clustering results for the 2015 incidence of Lyme disease. Describe any differences in the pattern of spatial clustering between this year and the average incidence.

4. Repeat step 2 with input as **Inciden_16** (name it **Lyme_incidence_Local2016**).

Section 2.1 Task 3: Provide a map of the clustering results for the 2016 incidence of Lyme disease. Describe any differences in the pattern of spatial clustering between this year and the average incidence as well as any differences between this year and 2015.

5. Repeat step 2 with input as **Inciden_17** (name it **Lyme_incidence_Local2017**).

Section 2.1 Task 4: Provide a map of the clustering results for the 2017 incidence of Lyme disease. Describe any differences in the pattern of spatial clustering between this year and the average incidence as well as any differences between this year and the two years before.

Section 2.2: Calculate spatial autocorrelation using Getis-Ord Gi*

For the first part of this section, you will calculate Getis-Ord Gi to determine if the data is clustered and the cluster's pattern type.

1. In the Geoprocessing toolbox on the left side of your ArcGIS Pro environment, navigate to **Spatial Statistics Tools ->Mapping Clusters->Hot Spot Analysis (Getis-Ord Gi*).**

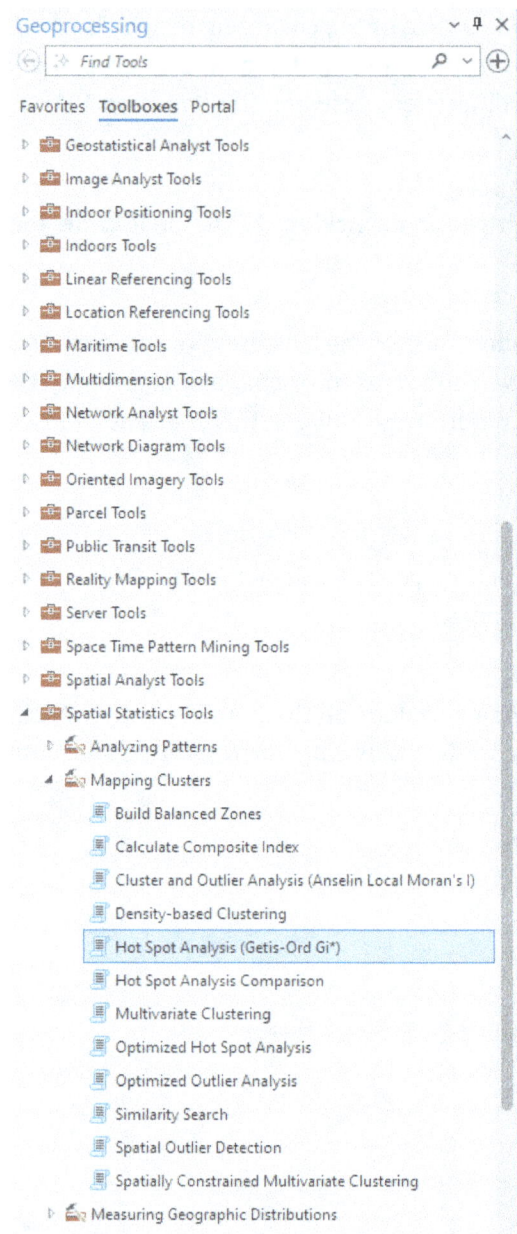

2. Set the input feature class to **Lyme_incidence** and the input field to **AverageInc** (Average incidence rates for the years 2000-2017). Click the folder next to the Output Feature Class field input box. Navigate to the **Lyme_disease.gdb** (you should already be there) and name the output feature class **Lyme_incidence_GstarAverage.** Confirm that the conceptualization of distance is set to **Contiguity edges only**. Leave all other fields blank and click **Run**.

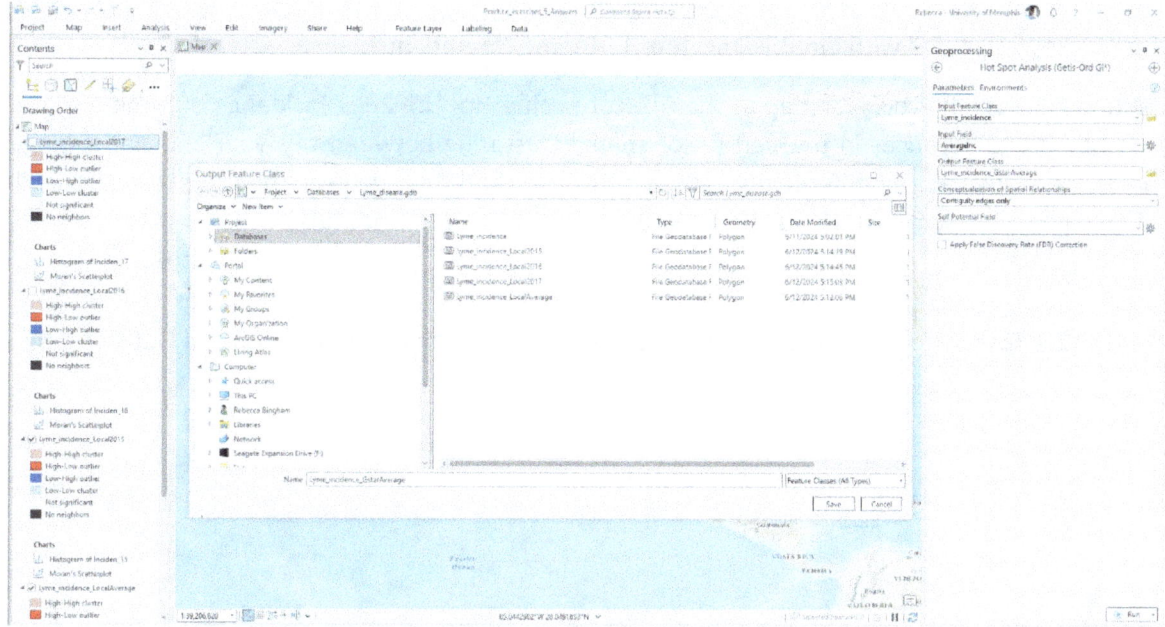

Section 2.2 Task 1: Provide a map of the GiZscore values for the average incidence of Lyme disease. Describe the pattern of spatial clustering including which areas exhibit clustering and which do not. (Note: you will have to go to the symbology and change the Field to **GiZScore Contiguity**).

3. Repeat step 2 with input as **Inciden_15** (name it **Lyme_incidence_Gstar2015**).

Section 2.2 Task 2: Provide a map of the GiZscore values for the 2015 incidence of Lyme disease. Describe any differences in the pattern of spatial clustering between this year and the average incidence. (Note: you will have to go to the symbology and change the Field to **GiZScore Contiguity**).

4. Repeat step 2 with input as **Inciden_16** (name it **Lyme_incidence_Gstar2016**).

Section 2.2 Task 3: Provide a map of the GiZscore values for the 2016 incidence of Lyme disease. Describe any differences in the pattern of spatial clustering between this year and the average incidence as well as any differences between this year and 2015. (Note: you will have to go to the symbology and change the Field to **GiZScore Contiguity**).

143

5. Repeat step 2 with input as **Inciden_17** (name it **Lyme_incidence_Gstar2017**).

Section 2.2 Task 4: Provide a map of the GiZscore values for the 2017 incidence of Lyme disease. Describe any differences in the pattern of spatial clustering between this year and the average incidence as well as any differences between this year and the two years before. (Note: you will have to go to the symbology and change the Field to **GiZScore Contiguity**).

Section 2.2 Task 5: Compare your results obtained from Anselin Local Moran's I and Getis-Ord Gi*.

CHAPTER 6. CHILDREN'S LEAD POISONING

Mastering the Concepts

This chapter explores the application of Geographic Information Systems (GIS) in examining children's lead poisoning. It highlights the methods that have proven most effective in analyzing the spatial epidemiology of childhood lead exposure. By leveraging GIS, researchers can identify high-risk areas, uncover demographic and geographic disparities, and guide targeted interventions.

GIS has been applied in various stages of lead poisoning interventions, from data preparation to multivariate mapping of blood lead levels (BLLs) and their risk factors, as well as spatial and statistical analysis. Address geocoding is commonly used to convert tabular data, such as children's addresses, into GIS-compatible formats. GIS functions have supported multivariate mapping, including linking socioeconomic data to screening records, map overlays, distance calculations, and visualizing lead exposure near demolition sites. More advanced spatial methods, such as clustering, autocorrelation, spatial regression, and risk modeling, have enhanced understanding of lead exposure patterns and risks. GIS-based studies are needed to improve surveillance, risk analysis, visualization of lead exposure, and geographically targeted community intervention strategies.

In the U.S., states collect child-specific data on elevated blood lead levels (EBLLs), including follow-up on medical treatment, environmental investigations, and lead exposure sources. At-risk children are screened via blood tests, with laboratories reporting all results to state health departments. States set reporting thresholds and required data elements, which are then shared with the CDC's national surveillance system. The CDC's latest blood lead reference level is 3.5 µg/dL, updated from the previous reference level of 5 µg/dL. This change reflects improved understanding of lead toxicity, emphasizing that even lower levels of lead exposure can harm children and warrant case management. Geographic and demographic disparities reveal that lead poisoning disproportionately affects children in low-income countries, marginalized populations, and those in lead-polluted areas. In the U.S., factors such as living in impoverished neighborhoods, pre-1950 housing, urban/rural status, race/ethnicity, socioeconomic status, population density, housing value, and nutritional status all independently heighten the risk of lead poisoning in children.

State health departments use surveillance systems to track childhood lead poisoning data. For instance, Tennessee's LeadTrack software automates patient data transfer and geocodes residential addresses. Some counties use local systems to manage remediation efforts, such as abated housing data. Geocoded coordinates enable GIS mapping of lead prevalence and associated environmental and socioeconomic risk factors, supporting targeted public health interventions. Spatial patterns guide community outreach and prevention.

Geocoding is essential for data aggregation and analysis in childhood lead poisoning research, linking socioeconomic and environmental factors to lead data while ensuring confidentiality. Integrating GIS into the data collection process could improve address accuracy, reduce spatial bias, and enhance screening practices, increasing case identification and data completeness.

Screening programs identify childhood lead poisoning early for treatment and prevention. Screening types include voluntary (less effective), mandatory (more comprehensive), and targeted (risk-based). Universal screening covers all children in a category, while targeted screening focuses on high-risk children. GIS-assisted strategies can improve screening policies to ensure appropriate prevention and reduce unnecessary testing.

The deduplication process removes duplicate or sequential test results from blood lead data collected by various laboratories. States maintain databases to identify and eliminate duplicates, using either the first or highest test result, or sometimes an average, for each child.

Many health departments shifted their approach from universal screening to targeted screening based on census parameters of median income, number of family members in the household, and previous lead poisoned children's locations to identify high-risk areas. Delineation of these high-risk areas with GIS helps local public health officials to do door-to-door testing, screening, and educational activities at local organizations, churches, and community centers.

GIS applications in examining lead poisoning provide essential insights into the magnitude and distribution of lead exposure within populations. Surveillance and screening efforts prioritize targeting "at-risk" groups, an approach that GIS can enhance by addressing spatial bias and disparities in reporting. Integrating GIS into health department screening practices can improve data accuracy and eliminate geographic inequalities in lead exposure assessments.

The use of finer geographic units, such as census block groups, is essential for achieving a more accurate analysis of lead poisoning patterns. Broader aggregations, like ZIP codes or census tracts, often overlook important details in population distributions. Longitudinal analyses can be complicated by changes in census boundaries over time. GIS-based methods can help address these challenges, ensuring confidentiality while accurately identifying and mapping high-risk areas.

Environmental studies show correlations between blood lead levels (BLLs) and African American populations, though few explore individual characteristics or broader socioeconomic and cultural factors. Incorporating these elements into GIS-based research could deepen understanding of lead poisoning's effects and inform targeted interventions.

Additionally, studies must account for intervention measures, such as housing abatement efforts, to avoid misinterpretation of BLL data. Addressing these gaps in GIS research would enhance awareness of lead poisoning's complexity and its impacts on vulnerable populations.

Mastering the Skills

Exercise 6. Spatial statistics of Health Data in ArcGIS Pro – Correlation of values to understand how one variable can impact another.

In this tutorial you are going to become familiar with correlation techniques to how demographic data can impact a child's exposure to lead.

OBJECTIVES

- Utilize exploratory data analysis techniques to search for relationships between demographics and children's lead exposure
- Utilize a widely used method for correlation analyses (Ordinary Least Squares: OLS)

Required data:

The original data sources listed below are provided for reference only. You do not need to download or curate these datasets from the original sources. All data, preconfigured GIS project files, and geodatabases required for this exercise are already included with the tutorial data and are available through the Resources page of the author's official website: www.esraozdenerl/resources. Instructions for accessing the tutorial data can be found in the Tutorial Data section of the Preface.

Original data sources:

1. Data from the Practice Exercises 6 folder. Sources:
   a. CDC Lead by Counties 2015 (www.cdc.gov)
   b. Demographics data (https://data.census.gov)

Section 1: Univariate Ordinary Least Squares Regression Analysis

Understanding how a variable can impact another variable, e.g., how certain demographics may affect the risk of exposure to lead for children, you can examine the relationship between demographic data and elevated blood lead levels (EBLL) using the OLS regression tool. The Ordinary Least Squares (OLS) tool conducts global regression analysis to predict a dependent variable based on its relationships with a set of explanatory variables.

Section 1.1: Running the Ordinary Least Squares Regression (OLS) tool

For the first part of this tutorial, you will use the OLS tool to determine key explanatory variables for your regression model.

2. Open the **Childrens_Lead.aprx** map file by navigating to the Practice Exercises 5 folder and double-clicking the file.

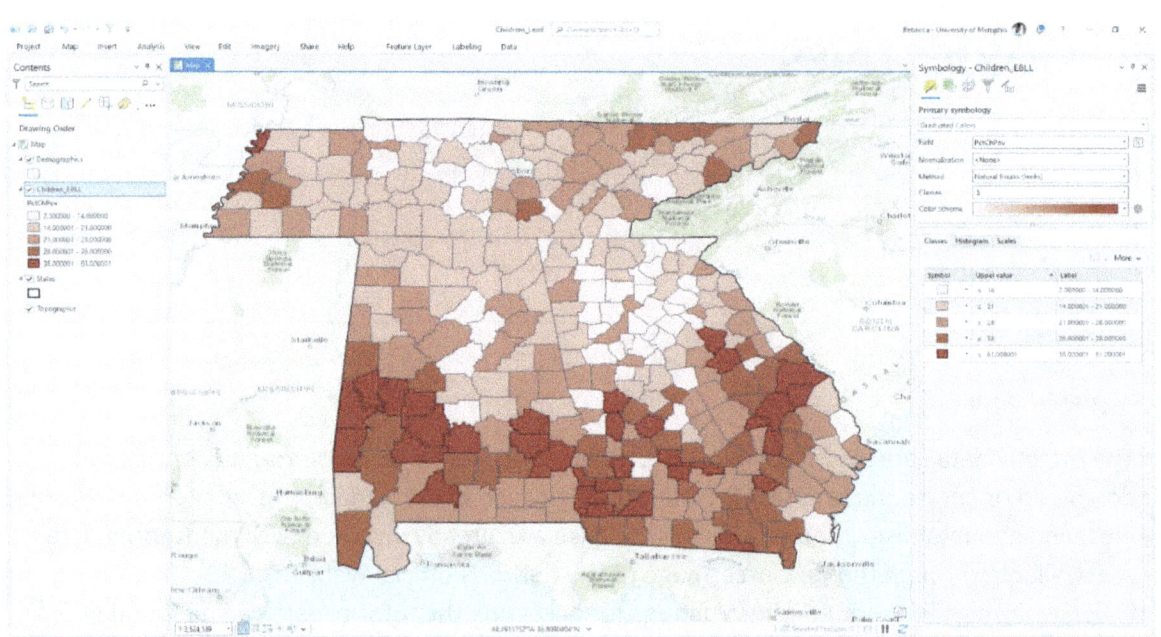

3. Next, you will have to set the data source for each layer in the map file as you did in Exercise 2. A short example of steps is included below. Do this for each layer.

a. Right-click the **Demographics** layer->Choose Properties->Select Data->Click Set Data Source
   b. Navigate to the **Childrens_Lead.gdb** geodatabase in the data folder for this exercise
   c. Choose the **Demographics** layer
   d. Click **OK**

4. Join the **Demographics** layer with the **Children_EBLL** layer by right-clicking the **Children_EBLL** layer->Joins and Relates->Add Join.

Add Join

Input Table
Children_EBLL

⚠ **Input Field**
FIPS

Join Table
Demographics

Join Field
FIPS

☑ Keep all input records
☐ Index join fields

Join Operation

[Validate Join]

[OK]

5. Next, you will want to click **Analysis** in the top panel of the ArcGIS Pro environment. Click **Tools** to open the Geoprocessing toolbox on the left side of your ArcGIS Pro environment. In the search bar, type **ordinary** and click **Ordinary Least Squares (OLS)**.

6. Set the input feature class to **Children_EBLL** and the Unique ID Field to **ObjectID [Children_EBLL.ObjectID_1]**. In the Output Feature Class box, type **OLSChildrenEBLL_Results**. Set the Dependent Variable to **ChLead** and check **AGE_UNDER5** in the Explanatory Variables. Scroll down and under Output Report File, navigate to your folder and save the report as **OutChLeadUnder5** (it will be in a pdf format). Click on Additional outputs to expand it, under Coefficient Output Table (optional), navigate to **Childrens_Lead.gdb** and save the table as **CoeffChLeadUnder5** (contains model coefficients, standardized coefficients, standard errors, and probabilities for each explanatory variable). Then, Diagnostic Output Table: navigate to **Childrens_Lead.gdb** and save the table as **DiagChLeadUnder5** (contains model summary diagnostics) Click **Run**.

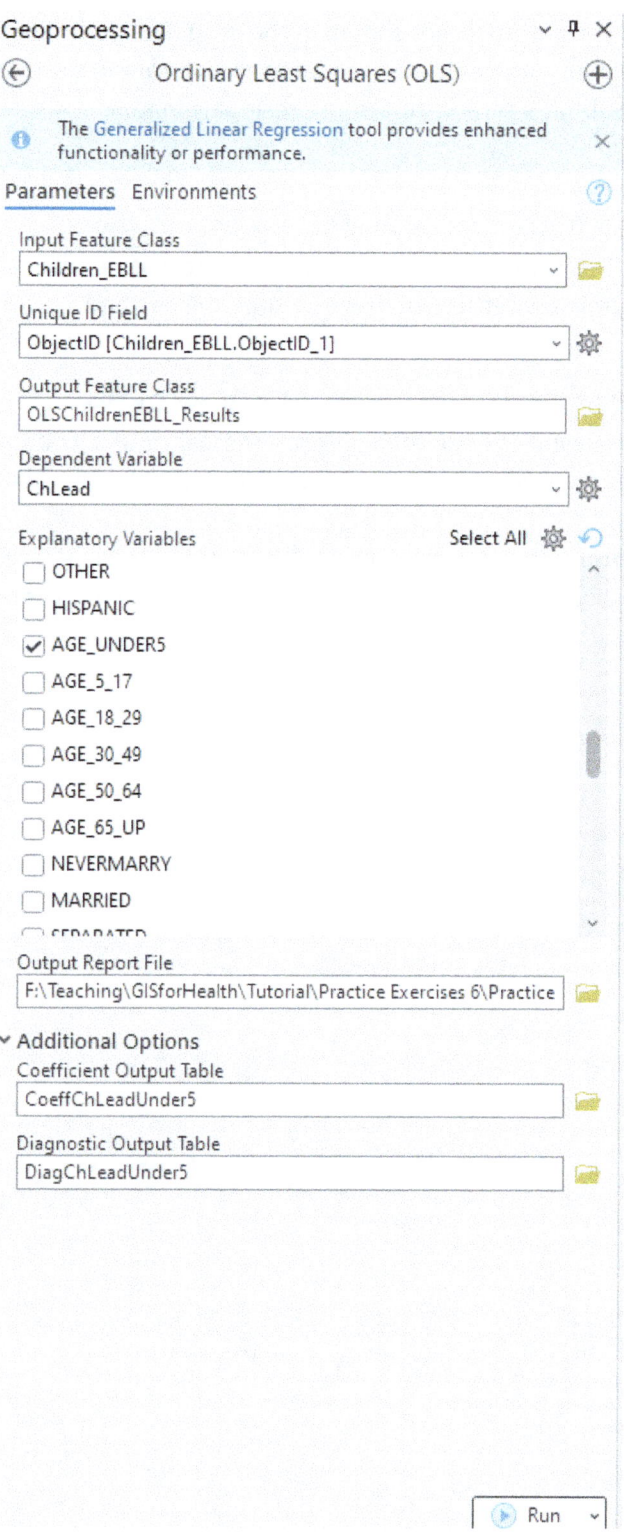

Section 1.1 Task 1: The residuals are shown in the map. Residuals are the unexplained portion of the dependent variable; some of the dependent variable remains unexplained because you tested only one variable. Submit a screenshot of the residuals map.

Section 1.2: Perform six checks to create a proper model

For the first part of this tutorial, you will use the Geoprocessing results window to perform the six OLS checks to determine whether you have a properly specified model.

1. From the Geoprocessing menu, click **View Details** and expand the Ordinary Least Squares results window. Right-click Messages and choose View. You will utilize the data in this report to conduct the six OLS checks and verify if the model is correctly specified.

   Note: You might observe a warning in the report regarding spatial autocorrelation after executing OLS. You will receive more information regarding this warning later on in the tutorial.

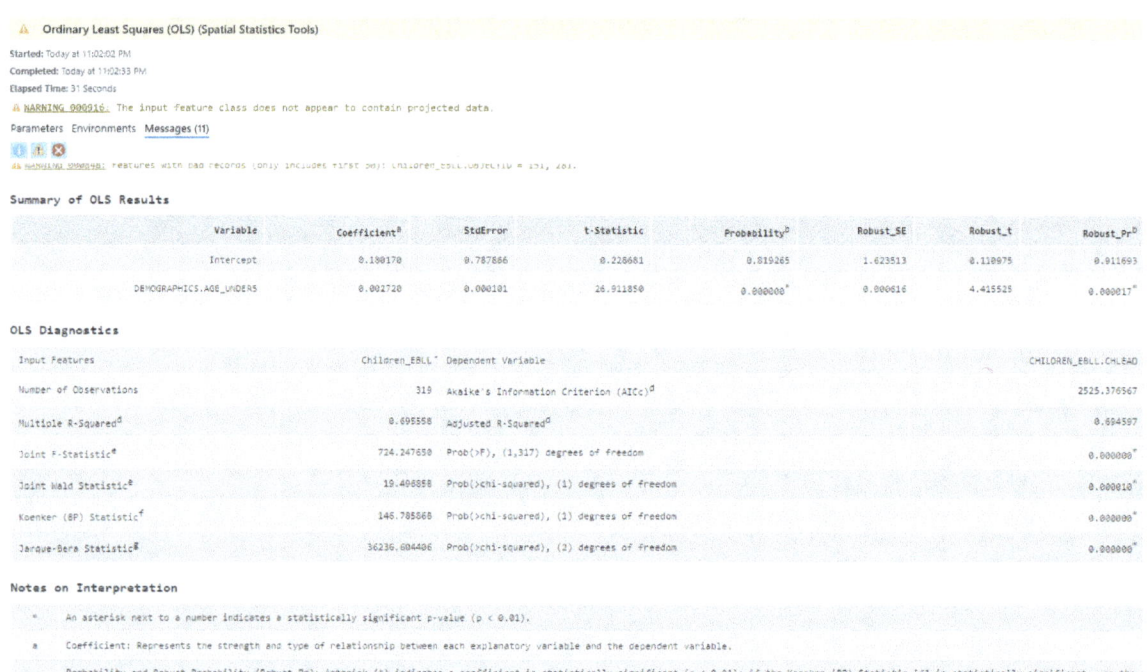

2. Check 1: Are the independent (explanatory) variables helping your model? In the Messages window, under Summary of OLS Results, locate the **AGE_UNDER5** variable and its coefficient.

Section 1.2 Task 1: Is it significantly different? What is the associated p-value (look at both Probability and Robust Probability, use the Koenker (BP) Statistic criterion to decide which one to use (Hint: If it is statistically significant, use the Robust Probability column (Robust_Pr) to determine coefficient significance).

3. Besides checking whether a variable is significant within your model, you also want to check the sign (+/-) associated with each coefficient. In the Messages window, locate the **AGE_UNDER5** coefficient.

Section 1.2 Task 2: What is the sign for the AGE_UNDER5 coefficient?

4. Are any of the explanatory variables redundant? When selecting variables to explain a phenomenon, it's important to choose variables that align with each other. The VIF diagnostic tool helps identify if any variables are redundant. Generally, a VIF over 7.5 indicates redundancy, and you should remove those variables one at a time until the redundancy is resolved. Since you ran OLS with only one variable, there is no VIF in this case. You'll encounter VIF in the next step when you create a multivariate regression model.
5. Is the model biased? The Jarque-Bera test evaluates whether the residuals (the difference between the observed dependent variable values and the predicted/estimated values) follow a normal distribution. If the Jarque-Bera diagnostic is statistically significant ($p < 0.01$), it indicates that the model is biased or skewed, meaning the residuals are not normally distributed. Normally distributed residuals suggest a properly specified model. In the Messages window, find the Jarque-Bera statistic and identify the statistically significant probability. An asterisk indicates that the residuals are biased for the population.

Section 1.2 Task 3: Based on the Jarque-Bera statistic, are the residuals normally distributed?

6. Do you have all key explanatory variables? Including the appropriate variables is crucial for a properly specified model. In this instance, you only used one explanatory variable, so additional variables are needed. Omitting explanatory variables results in

misspecification, a common issue in regression models that can cause failures in one or more of the six checks. Consequently, results from a misspecified OLS model are unreliable. If you scroll to the bottom of the report, you will see a warning stating that you should run Spatial Autocorrelation to ensure that the residuals are not spatially autocorrelated. Any structure in the residuals indicates that you do not have all the key explanatory variables. Close the Messages window.

7. In the Search window, type spatial autocorrelation and press Enter. Open the Spatial Autocorrelation (Global Moran's I) tool. Enter the following parameters:
    - Input Feature Class: **OLSChildrenEBLL_Results**
    - Input Field: **Residual**
    - Check box to **Generate Report** to create a graphical summary of results as an HTML file.
8. Click **Run**.

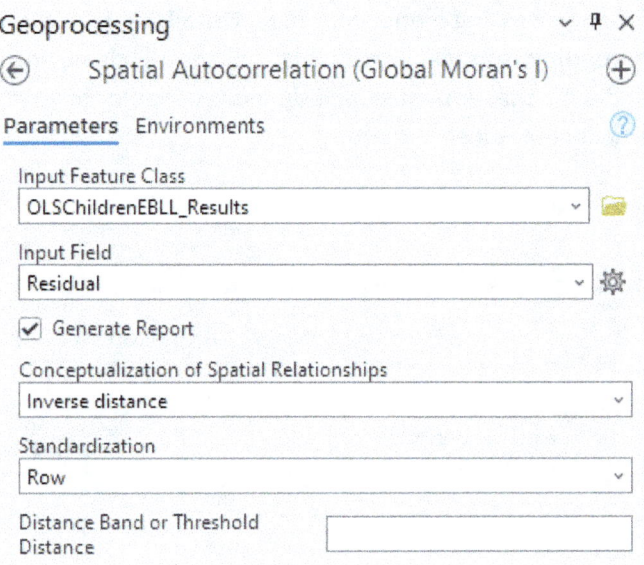

9. From the Geoprocessing menu, click **View Details** and double-click the report file. The result of running spatial autocorrelation shows clustering in the residuals. This clustering means that you must include more variables in your analysis, which you will do in the next step. Close the report.

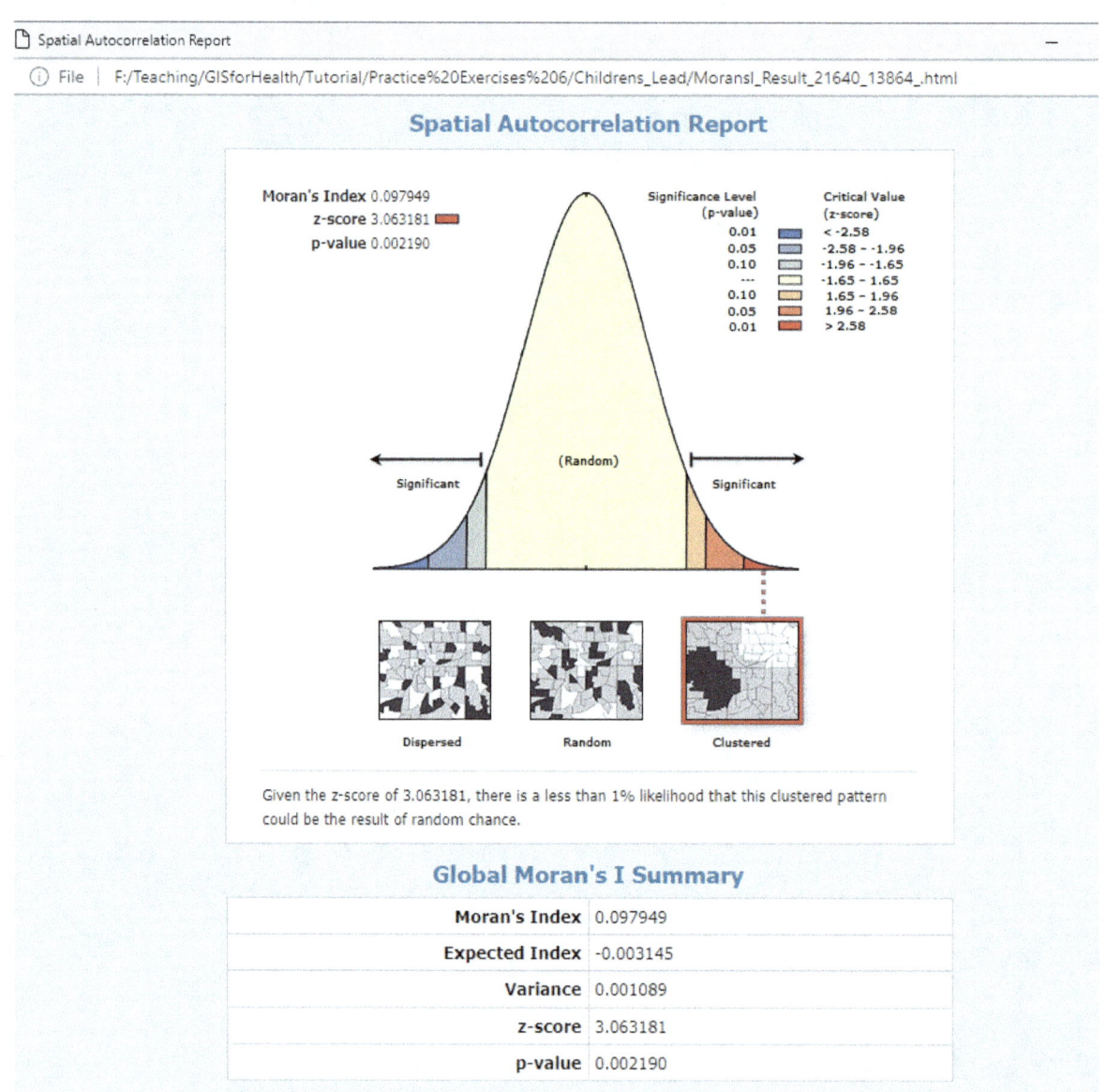

10. This final check assesses model performance by examining two values: Adjusted R-Squared and AIC. The Adjusted R-Squared value is crucial, but it should not be relied upon alone unless the other checks are passed. Adjusted R-Squared values range from 0 to 1, representing the percentage of how well your dependent variable is explained by the model. Navigate to your work folder and open **OutChLeadUnder5.pdf**, then scroll to OLS Diagnostics.

OLS Diagnostics

Input Features	Children_EBLL	Dependent Variable	CHILDREN_EBLL.CH LEAD
Number of Observations	319	Akaike's Information Criterion (AICc)['d']	2525.376567
Multiple R-Squared['d']	0.695558	Adjusted R-Squared['d']	0.694597
Joint F-Statistic['e']	724.247650	Prob(>F), (1,317) degrees of freedom	0.000000*
Joint Wald Statistic['e']	19.496858	Prob(>chi-squared), (1) degrees of freedom	0.000010*
Koenker (BP) Statistic['f']	146.785868	Prob(>chi-squared), (1) degrees of freedom	0.000000*
Jarque-Bera Statistic['g']	36236.604406	Prob(>chi-squared), (2) degrees of freedom	0.000000*

11. Locate Adjusted R-Squared. An Adjusted R-Squared value of 0.69 indicates that Children under the age of 5 alone does not tell the children's lead story (it explains less than a percent of EBLL in children). For a properly specified model, you want to explain more of the dependent variable. Adding other explanatory variables might help.
12. The next statistic that evaluates model performance is Akaike's Information Criterion (AIC). It is located above the Adjusted R-Squared value. Notice that the AIC value is 2525.376. AIC measures how well a statistical model fits the data. It is not a hypothesis test but a tool for model selection. When comparing several models for the same dataset, the model with the lowest AIC is considered the best. The AIC value will be compared to the AIC of the next model you create for the same analysis. AIC is always relative; it doesn't have an absolute high or low value but is meaningful only when compared to another model testing the same dependent variable.
13. After performing the six checks on the OLS results, you can see that this model is not properly specified, and more explanatory variables are needed.
14. Close the Messages and Results window. Close the OLS Diagnostics

Section 2: Multivariate Regression model

Now that you've determined that using only children under the age of 5 as an explanatory variable doesn't provide a properly specified model, you will use multiple variables to explain EBLL in children.

Section 2.1: Running the Ordinary Least Squares Regression (OLS) tool

For the this part of this tutorial, you will use the OLS tool to determine key explanatory variables for your regression model.

1. Click **Analysis** in the top panel of the ArcGIS Pro environment. Click **Tools** to open the Geoprocessing toolbox on the left side of your ArcGIS Pro environment. In the search bar, type **ordinary** and click **Ordinary Least Squares (OLS).**

2. Set the input feature class to **Children_EBLL** and the Unique ID Field to **ObjectID [Children_EBLL.ObjectID_1]**. In the Output Feature Class box, type **OLSChildrenEBLLResults_Full**. Set the Dependent Variable to **ChLead** and check **WHITE, BLACK, AMERI_ES, ASIAN_PI, OTHER, HISPANIC, AGE_UNDER5, AGE_5_17, MARRIED, SEPARATED, WIDOWED,** and **DIVORCED** in the Explanatory Variables. Click **Run**.

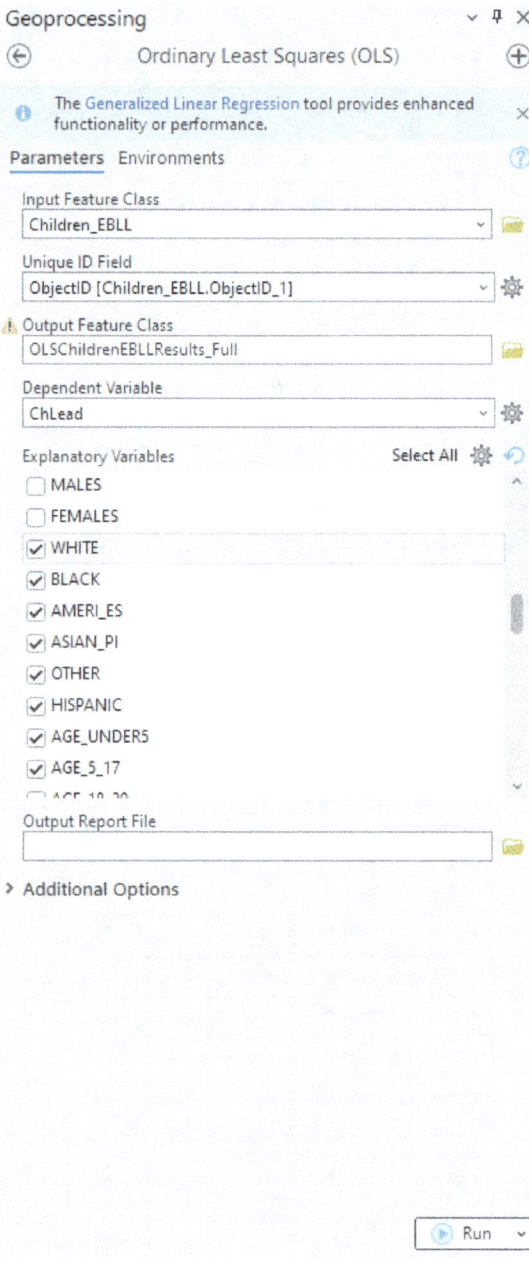

Section 2.1 Task 1: Submit a screenshot of the residuals map.

Section 2.2: Perform six checks to create a proper model

For the first part of this tutorial, you will use the Geoprocessing results window to perform the six OLS checks to determine whether you have a properly specified model.

1. From the Geoprocessing menu, click **View Details** and expand the Ordinary Least Squares results window. Right-click Messages and choose View. You will utilize the data in this report to conduct the six OLS checks and verify if the model is correctly specified.

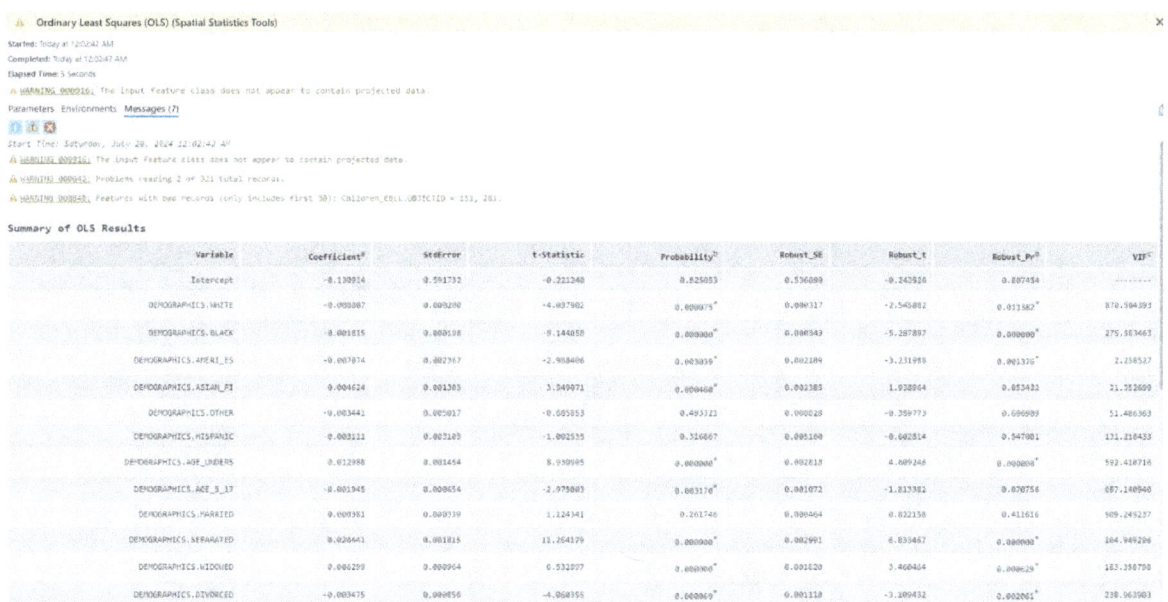

Section 2.2 Task 1: Please answer the following questions:

- Are the probabilities statistically significant?
- Are any of the explanatory variables redundant? (Hint: Look for a VIF value over 7.5.)
- What are the Adjusted R-Squared and AIC values?
- What do these values tell you about your model?
- Is the result of the Koenker test statistically significant, and what does this mean?
- Is the result of the Jarque-Bera test statistically significant?

You can run spatial autocorrelation even that the Jarque-Bera test was not statistically significant.

2. In the Search window, type spatial autocorrelation and press Enter. Open the Spatial Autocorrelation (Global Moran's I) tool. Enter the following parameters:
    - Input Feature Class: **OLSChildrenEBLL_FULL**
    - Input Field: **Residual**
    - Check box to **Generate Report** to create a graphical summary of results as an HTML file.
3. Click **Run**.

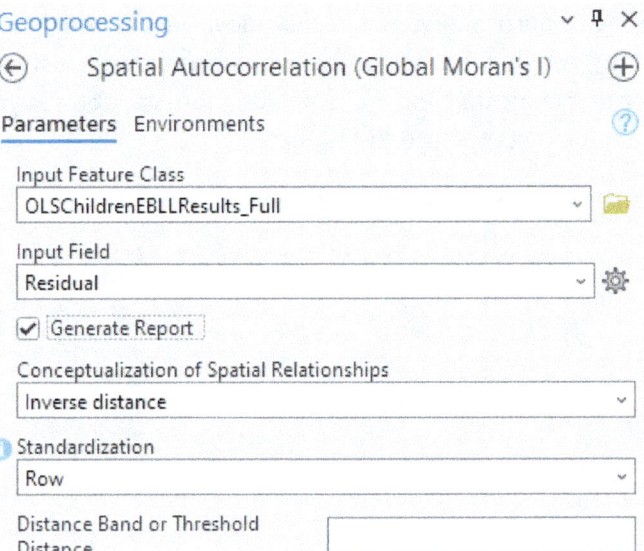

4. From the Geoprocessing menu, click **View Details** and double-click the report file. The result of running spatial autocorrelation shows clustering in the residuals. This clustering means that you must include more variables in your analysis, which you will do in the next step. Close the report.

Section 2.1 Task 2: Submit a screenshot of the report.

CHAPTER 7: ONLINE GIS APPLICATIONS

Mastering the Concepts

Geographic Information Systems (GIS) have undergone a remarkable transformation over the past two decades, evolving from specialized tools used primarily by experts into accessible platforms that empower a broad range of users. The internet has played a crucial role in this shift, making GIS-based dashboards and mapping portals widely available to both health professionals and the general public. This transition has been especially impactful in the context of public health, where the need for real-time, data-driven decision-making is more critical than ever.

Previously, GIS was largely confined to desktops and complex software that required specialized knowledge to operate. Today, with the rise of cloud computing, improved broadband infrastructure, and mobile technology, GIS has become a widely accessible tool. Web-based GIS platforms have transformed how users interact with geographic data, allowing them to easily access, analyze, and share maps and visualizations without the need for advanced training. These platforms often feature user-friendly interfaces with simple dashboards that present complex data in a way that is easy to understand, enabling people with little to no GIS experience to make informed decisions.

In the realm of public health, these GIS tools have proven invaluable in responding to health crises. The COVID-19 pandemic is a prime example of how internet-based GIS applications can provide the public with the real-time data they need to understand the progression of a health threat. From the earliest days of the pandemic, people around the world turned to maps and dashboards powered by GIS to track the virus's spread, its impact on different regions, and the availability of healthcare resources. The ability to visualize geographic data helped to contextualize raw case numbers, showing not only how the virus spread geographically but also offering insights into how local factors, such as population density and healthcare access, influenced the severity of outbreaks.

One of the most widely recognized GIS tools during the pandemic was the interactive dashboard developed by Johns Hopkins University's Center for Systems Science and Engineering (CSSE). Launched in January 2020, the CSSE dashboard became a go-to resource for millions of people around the world. It displayed live updates on confirmed cases, deaths, and recoveries, while interactive maps showed the geographical distribution of the virus. The dashboard's success was due, in part, to its integration with multiple authoritative data sources, including the World Health Organization (WHO) and the U.S. Centers for Disease Control and Prevention (CDC), which ensured the information was up-to-date and accurate. This seamless integration of data allowed users to track the pandemic's progression in real time, giving policymakers and health officials the tools they needed to respond quickly and effectively.

Esri's ArcGIS Living Atlas, which supported the CSSE dashboard, played a critical role in disseminating data during the pandemic. ArcGIS made it possible for organizations around the world to easily share and access up-to-date information. This global data-sharing network helped health professionals and researchers create their own dashboards and apps, allowing them to tailor their analysis to specific regions and populations. As a result, GIS-based platforms became integral to managing the global health crisis, providing valuable insights into trends, demographics, and healthcare system performance.

The success of GIS in public health during the pandemic has paved the way for new applications in digital health. As technology continues to evolve, mobile health applications (mHealth apps) have emerged as powerful tools for managing individual and public health. These apps, many of which incorporate GIS features, allow users to track their health, monitor outbreaks, and access real-time data about health risks in their area. GIS technology provides the geographic context needed to interpret health data, showing users where healthcare resources are available, mapping pollution levels that affect respiratory conditions, or identifying areas with high rates of chronic illness.

In addition to tracking infectious diseases, mHealth apps now allow users to monitor a range of health conditions, from chronic illnesses like diabetes and asthma to mental health concerns. By incorporating geographic data, these apps help individuals understand how their environment impacts their health. For example, they can show users where they might be at higher risk due to environmental factors or help them find healthcare providers in their area. As more people rely on these apps, GIS is becoming an essential part of managing personal health and wellness.

Moreover, the growing use of GIS in public health extends beyond personal health management to address broader social determinants of health. By integrating environmental data—such as air quality, water contamination, and access to healthy food—GIS can identify at-risk populations and inform targeted interventions to reduce health disparities. This geographic approach helps public health agencies better understand how different factors contribute to health outcomes and design programs that address these challenges.

Online GIS-based mapping portals are revolutionizing the way health organizations and systems approach data visualization and decision-making. These platforms integrate diverse datasets, allowing users to analyze complex health information in an intuitive and accessible format. By combining layers of health outcomes, social determinants of health (SDOH), and community resources, these portals empower stakeholders to identify disparities, allocate resources effectively, and develop data-driven strategies to improve public health outcomes. The ability to visualize geographic trends in health data fosters greater collaboration among health professionals, policymakers, and communities, driving more equitable and impactful interventions.

The Tennessee Heart Health Network Mapping Portal exemplifies this innovation by providing an interactive platform that integrates cardiovascular health outcomes, such as hypertension and high cholesterol, with community resources like health clinics and SDOH layers. This comprehensive tool offers invaluable insights into the interplay between health, environment, and accessibility, supporting targeted interventions to reduce disparities and improve heart health equity across the state.

The future of GIS in public health looks even more promising with the rise of crowdsourced health data. Some mobile apps allow users to contribute their own health information, such as symptoms, vaccination status, or even observations of local outbreaks. By collecting data from a wide range of individuals, these platforms can provide real-time insights into disease trends and public health risks. This crowdsourced data, when combined with traditional health surveillance systems, offers a more granular understanding of health dynamics and can help public health authorities respond more quickly and efficiently.

The digital transformation of public health through GIS is just beginning, and its potential is vast. As more health data becomes accessible and mobile apps continue to evolve, GIS will play a central role in how we understand and respond to health threats. By providing real-time insights, supporting decision-making, and empowering individuals to manage their health, GIS is reshaping the future of public health. Whether through web portals, mobile apps, or crowdsourced data, GIS is making it possible for more people to engage with health information and contribute to solving global health challenges. As these tools continue to evolve, they will undoubtedly play an even greater role in improving health outcomes worldwide.

Mastering the Skills

Exercise 7. Constructing a Simple ArcGIS Experience Builder Site and Configuring Multiple Widget Types

With the use of this tutorial, you will be able to construct a website using Experience Builder involving some of the available widgets.

> **OBJECTIVES**
> - Create a basic Experience Builder site using a webmap containing census tract level social determinants of health data from the CDC
> - Utilize a variety of Experience Builder widgets to portray the data, while providing a useful design experience to the student in the process

Requirements;

1. An ArcGIS project file from Practice Exercises 7 folder titled "Experience_Builder_Tutorial" should be opened within ArcGIS Pro, and then shared as a webmap to your ArcGIS Online account with the title "Experience_Builder_Tutorial_<YOUR INITIALS>".
   - In the upper menu, click Share → Web Map.
   - When prompted, share the map with your ArcGIS Online account.
   - Use the following title format:
     Experience Builder Tutorial_YourInitials *(Example: Experience Builder Tutorial_EA)*
   - Before sharing, always click "Analyze" to check for any errors.

   If you see errors:

   - Click the three dots next to the error message. ArcGIS will display prompts and guidance to help you fix them.
   - To resolve the 'unique numeric IDs are not assigned' error, select 'open Map Properties to allow assignment', and select the 'Auto-assign IDs sequentially'.

- To fix a coordinate system error, select update the 'map to use the Basemap's coordinate system'.
- For a CBO layer error, select 'make the Location variable visible', then save the layer by clicking the *Save* button in the upper menu.

After fixing all issues, click Analyze again. Once you see all green check marks, your web map is ready to share.

2. Original Data Sources:
    b. Tennessee_CensusTracts_SDOH layer is from CDC\ATSDR\Office of Innovation and Analytics\Geospatial Research, Analysis, and Services Program (GRASP) and was downloaded from ArcGIS ONLINE and filtered for TN tracts only.
    c. CBO (Community Based Organizations) layer is 64 point features, primarily in the Shelby County, Tennessee area, representing community-based organizations from a variety of categories.

Section 1: Initial Site Construction

ArcGIS Experience Builder is a tool designed to create web apps and pages using data, maps, and content with minimal or no coding required. It empowers its users to deliver responsive web experiences with little effort.

Section 1.1: Create the Experience Builder File

1. Sign-in to ArcGIS Online (AGOL). Click on the dot-grid next to the username in the upper right corner of the page to access the ESRI suite of apps. Select "Experience Builder."

2. From the webpage that opens, you can select from your existing experiences or create a new one. For the sake of this tutorial, click on "+Create new" in the upper right-hand corner.

3. The template page opens next. Many options are available that suit just about any purpose. Oftentimes, the "Blank Fullscreen" template is chosen. However, this option limits the vertical size of the page and the amount of content that can be included. Select the blue "Create" button associated with the "Blank scrolling" template (NOTE: feel free to mouse-over the template options to view descriptions of their intended use).

4. Your new Experience Builder file has been created (but not yet published). If the "Getting Started" tour window opens, feel free to proceed through it or to skip it. Click on the title "Untitled Experience 1" on the upper left corner of the page and change your Experience Builder site's title to whatever you desire. Notice that next to the title it says "Draft." This will remain the case until you publish the site later.

Section 1.2: Format the Page

1. On the left side of the webpage, you will see the following menu bar:

- Insert Widget
- Page
- Data
- Utility Services
- Theme
- General

2. Select the "Page" option. Notice there are now two tabs available with this selection, "Page" and "Window." Hover over the title of the only page listed, titled "Page." Click on the three option dots and select "Rename." Change the title to an appropriate name.

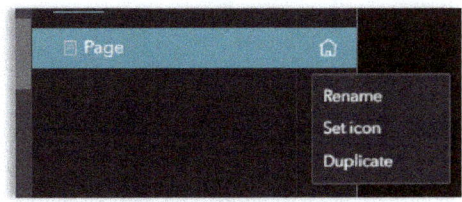

IMPORTANT NOTE: *When changes are made to your site, make sure to click on the "Save" icon near the upper right hand corner of the screen periodically to save your edits/changes. However, upon saving, the option to undo is reset. Even when saving, the site still is not "published."*

3. When a page, or window, or widget is selected in the left-side pane, a corresponding pane opens on the right-side, providing the opportunity to adjust data sources, formatting, and connections between content, depending on what is selected. With your Page selected in the left-side pane, toggle on both "Header" and "Footer."

4. You now have the opportunity to adjust the height and colors of both the header and footer. In the header group, change the height to 120 pixels. Because the default footer color is white, it might be difficult to realize it is toggled on. Change the color by clicking on the box to the right of "Fill" in the footer group of settings. For the time being, choose one of the Standard Colors that are presented. The background color of the page itself can be set as well. In the Page pane on the right-side, click the white

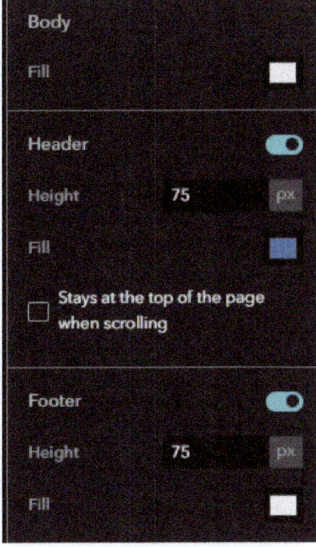

173

box next to "Fill" in the "Body" group. Try several different colors until the desired one is found.

Remember to save your work periodically.

Section 1.2 Task: Select "Theme" on the far-left toolbar.

Explore different themes.

> NOTE: You will be submitting your results of these "tasks" at the end of this tutorial by providing the URL of your published Experience Builder site.

Section 1.3: Image and Text Widgets

1. Later, you will be adding a WebMaps to your site that has social determinants of health data for each census tract in Tennessee. Let's customize the header in preparation for that. Whatever content is contained in the header on one page automatically appears on every page (the same is true with footers). Let's add an image and some text to the header. Hover your mouse over the header and click on "Edit Header." When content is added to any part of an Experience Builder site, it is done so using "widgets." Let's first add an Image Widget to the header.
2. On the menu bar on the far-left side of the site, select the "Insert Widget" option. By default, every widget that is available is shown. At the top-right of the pane, click on the search icon and type in "image." Using your mouse, click and drag the "Image" widget into the header area. This is simply a container; you must now add the image. Since the Image Widget is still selected, the image pane appears on the right-side of the page. Images can be displayed from a URL address, or an uploaded image. Click on "select an image."

Section 1.3 Task (a): For the sake of this exercise, find an open-source image related to health and store it locally on your computer. Click "Upload" in the "Select an Image" pane, navigate to your store image and select it. The image now appears in the Image Widget of the header.

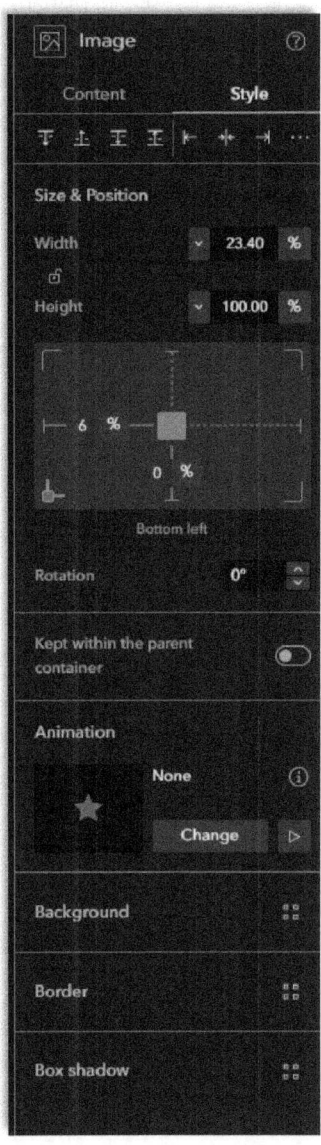

3. Each widget's pane provides many options to customize its appearance and usage. Click on the "Style" tab in the Image Widget pane.

4. The width and height of the widget can be controlled based on percentages and pixels. Click on the "%" sign next to the width value and select "px" for pixel. Vary the value for the width and see how it affects the size of the Image Widget. Do the same for the height value, until you are content with the size of the image in the header.

5. The positioning of widgets can be controlled similarly, using percentages based on the total screen size or the size of the container (more on this later) that holds them, or absolute pixel values. In the image below, the box in the center represents the widget. The "%" can be clicked on and changed to pixels; the numerical value can be typed in. Also, by clicking on any of the four vertical or horizontal lines, you can determine from where the positioning is referenced.

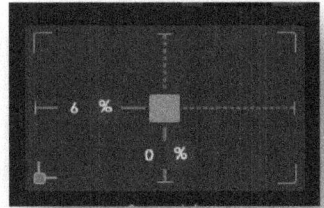

Section 1.3 Task (b): Change the positioning from the left edge to 65 pixels and determine the pixel value from the bottom edge that centers your image vertically.

6. Let's next add a descriptive text to the header. In the search box for the Insert Widget option on the left-side of the page, delete "image" and type in "text." Drag the Text Widget into the header area and position as desired. Double click the Text Widget box and type

in an appropriate title for a page that presents social determinant of health for Tennessee. Click outside of the Text Widget, but still within the header, and then click on the Text Widget once again.

Section 1.3 Task (c): Using the Text pane on the right-side of the page, adjust the font, font-size, etc. to your liking. Once again, click on the "Style" tab and adjust the size and positioning of the Text widget in the same fashion as you did the Image Widget.

7. Click in the center of the canvas area to exit header editing.

Remember to save your work periodically.

Section 2: Adding Content Using Widgets

Section 2.1: The Map and Search Widgets

You will now add Map and Search Widgets to display a webmap and find locations on it. However, before doing so, it is advisable to add some layout containers for the widgets. It is best practice to use items such as the "Row" and "Column" layout containers to better control the size and position of the elements of your page as the browser window changes size and proportion.

1. Go to the Insert Widget option on the toolbar on the far-left side of the page. Search for the "Row" widget. Drag the Row Widget onto the center of the canvas. It might be that this new item doesn't fill the width of the page. That is ok; it can be adjusted later.

2. Next, search for the "Map" widget. Drag the Map Widget into the Row that you previously put on the canvas. Using the sizing bars on the Map Widget, resize it to fill the entire Row Widget.

3. Currently, there is no map associated with the Map Widget. Let's change that. Previously, you uploaded the "Experience_Builder_Tutorial_<YOUR INITIALS>" webmap to AGOL. To display this map in the Map Widget, you must first make the data connection. On the toolbar on the far-left side of the page, select the "Data" option.

4. Click on the blue "+Add data." The available maps, scenes, and layers from your personal content and organization, along with ArcGIS Online and Living Atlas, should appear. Click on the "Experience Builder Tutorial" webmap and then select "done" in the bottom right-hand corner of the page. You are redirected back to the Experience Builder interface. Notice that the webmap still does not display in the Map Widget.

5. Click on the Map Widget and then direct your attention to the Map pane on the screen's far-right side. In the "Source" group of the "Content" tab, click on "Select Map."

6. Click on the Experience Builder Tutorial webmap's title in the "Select Data" panel that has opened.

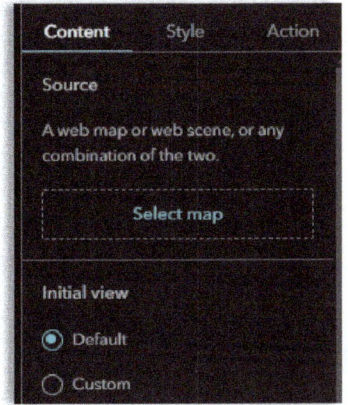

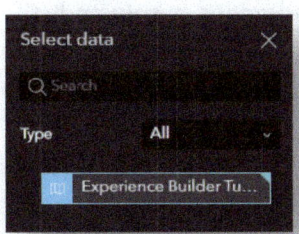

7. "X" out of the "Select Data" pane. In the "Map" pane, in the "initial view" grouping, select "custom" and then click "Modify."

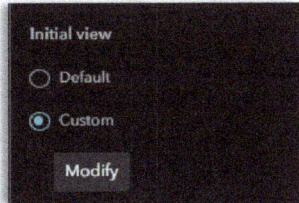

8. In the "Modify initial view" window that opens, use your mouse to determine the default view of your map. Click "OK" to close it. Save your work.

9. Let's get a good view of what's been done so far. To the immediate right of the save button is the "Preview" button. Click on this…a new browser tab opens with a preview of your site.

Section 2.1 Task (a): Having previewed your site, what changes to sizes and placements should be made? Oftentimes if percentage values are used for positioning, resizing the window changes appearances in a negative way. Edit your header so that both the Image Widget and the Text Widget have positions represented by pixels rather than percentages. "Save," "Preview," and resize the browser, confirming the results are as desired.

177

Section 2.1 Task (b): In the "Tools" grouping of the Map pane, there are many available options that can be toggled for your Map Widget. Explore each option and use what you deem appropriate (NOTE: toggle off the search option, we'll add that in a different way)

10. Further down in the "Options" grouping of the Map pane is a toggle for "Enable pop-up." Because you will display map data using several other widgets, toggle off the pop-ups.

Remember to save your work periodically.

11. Let's add a search box to locate positions on the map. In the Insert Widget tool on the far-left side of the page, search for "Search." Drag the widget directly onto the map. When this widget is selected, the Search pane appears on the far-right side of the page. In the Custom search sources section, click on the blue "+New search source" button in that pane.

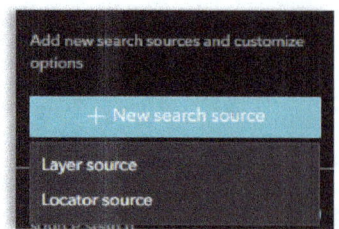

12. In the drop-down menu, you have the choice of "Layer source" (features from a given layer can be searched) and "Locator source" (geographic locations can be found). Choose "Locator source" and click "Select utility" in the "Set locator source" pane that appears. Choose the "ArcGIS World Geocoding Service" option. Leave all the defaults intact and close the "Set locator source" pane.

13. Click the "Style" tab in the "Search" pane and adjust the size and position of the widget.

14. Click the "Action" tab and then click on "Add a trigger."

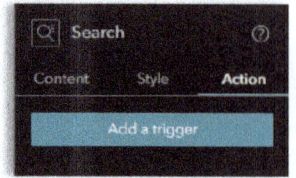

15. In the "Select a trigger" pane, click on "Record selection changes." Since we want interaction with the Search Widget and the Map Widget, under "Widgets" select "Map." Click on "Zoom to" so that a search selection causes the map to zoom to the search location. In the "Zoom scale" grouping, check the "Custom" box. Entering a value adjusts the scale of the map that results when zooming in upon selection. Put a value of your choice in this box.

16. In the "Search" pane, click "Add a trigger" again and select "Records created" in the "Select a trigger" pane. Click "Map" under "Widgets." Select "Show on map." This will

drop a marker on the webmap when you select a location from the Search Widget. Customize the symbol by clicking on the box to the right of the word "Point." Choose the symbology that you find appropriate.

17. Save your work and click on the "Preview" button. Test out the search widget with a location.

Section 2.1 Task (c): Adjust the value for the custom zoom so that when the map zooms in after a search selection, you are content with the map extent shown.

Section 2.2: The Text Widget and Dynamic Text

Each of the census tracts in the webmap has 9 attribute field corresponding to social determinants of health (SDOH). Let's add a text box that displays these attributes associated with a particular census tract when it is selected in the map.

1. From the Insert Widget option on the far-left side of the page, search for the Text Widget and drag it onto the canvas, to the right of the Map Widget within the Row Widget. Depending on the background color of your page, it might be difficult to view any text you include. In the Text pane on the far-right side, select the "Style" tab. Expand the "Background" section and select "Custom" at the bottom of the popup.

2. Next, click the color box to the right of "Fill" and choose white for the color. Double-click within the text widget and type a suitable title for the display of the SDOH. If you now highlight your title using your mouse, you can adjust the font size, color, etc.

3. Below the title, you will code the Text Widget to display data from a selected tract of the webmap. In the Text pane on the far-right side, select the "Content" tab and then toggle "Connect to data." Click on "Select data" and then choose the "Census tract" layer from within the "Select data" window that opened. So that data selected from the webmap is displayed, from the drop-down menu under Census tract in the Text pane, check the "Selected features" box.

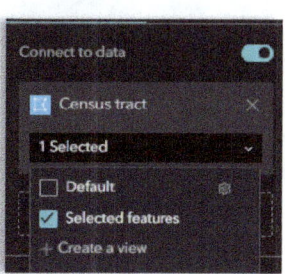

179

4. Next, click in the Text Widget and type "No high school diploma: " To get data from the selected census tract to display, you must access the "Dynamic content" feature above the Text Widget:

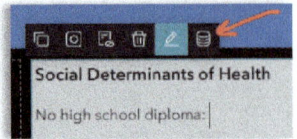

5. After clicking on the dynamic content option, the Dynamic content window opens, which has three tabs. While on the Attribute tab, you can either scroll down and click on "No high school diploma (%)," or start typing a few words like "no high" in the search bar and select the attribute from the results. Next, click on the attribute in the Text box:

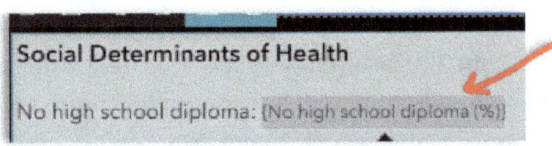

6. Then, select the Expression tab in the Dynamic content window. Click on the title box, delete the title that is there and replace it with an empty space using the spacebar (You've just set the title to an empty space, which will be displayed in the widget when there is no map selection). At the bottom of the window, select the tools gear. Toggle on "Number formatting," set the decimal box to 2, and the click "Insert." Close the Dynamic content window.

7. Save your work and preview the site (Note: click outside of the Text Widget before saving). In the preview, select one of the census tracts and observe that a value appears for the "No high school diploma (%)" in that census tract.

Section 2.2 Task: In the list of attributes for each census tract, there are a total of nine percentages. Repeat the above steps for the remaining eight SDOH. Then, format all of the content in the Text Widget, including the dynamic content, in an appropriate manner.

Section 2.3: The Chart Widget

1. To display the SDOH graphically, you will add a Chart Widget. From the Insert Widget option on the far-left side of the page, search for the Chart Widget. Drag a Chart Widget into the area below the Row Widget which contains the Text and Map Widgets. Using the sizing bars, increase the width of the Chart Widget to match the width of the page. In the Chart pane on the far-right side of the page, select "Select data." In the "Select data" window, navigate to and select the "Census tract" layer.

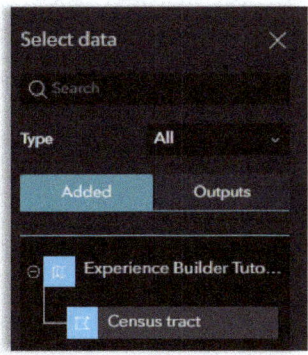

2. Back in the Chart pane, from the drop-down menu within the "Census tract" box, select "Selected features" (This will enable selections from the Map widget to have data charted). Still in the Chart pane, under "Chart type" click on "Select chart." Choose the first of the column chart types:

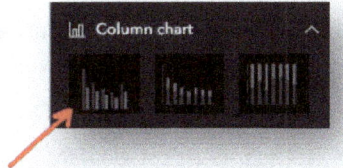

3. In the "Data" grouping, still on the Content tab, of the Chart pane, select the drop-down menu for "Category field" and choose "Census tract FIPS."

4. For the "Statistics" option, select "No aggregation" from the drop-down.

5. From the drop-down menu in the "Number Fields" option, check the boxes for the nine percentage SDOH attributes (NOTE: the order in which you check the boxes determines the order in which they are displayed!).

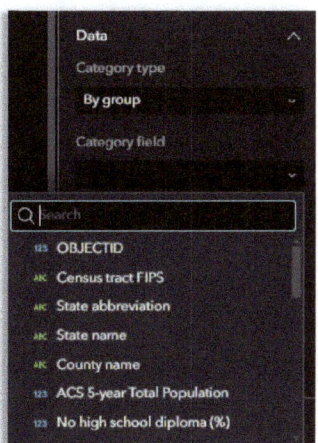

6. Next, the chart must be formatted. Expand the "Series" grouping in the Chart pane. Toggle on "Data label" (this places a value at the top of bar). Click the pencil next to "No high school diploma (%)."

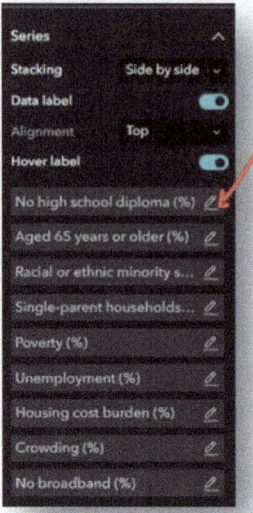

7. Edit the label to something shorter. This label will appear in the chart legend, as well as when the mouse hovers over a bar. If desired, the color of the bar can also be edited.

Section 2.3 Task: Edit the title of each of the nine SDOH.

8. Expand the "Axes" grouping. Select the edit pencil next to "X axis."

9. Give your x-axis a title. Repeat for the "Y axis."

10. Expand the "General" grouping. Give your chart an appropriate title. Also, toggle on "Legend." Explore the other "General" settings. For example, check "customize no data message" and change the wording to "No data available yet. Click on a census tract on the map to populate the chart." Since your chart depends on user selection (like a census tract), the new message should guide the user to take action.

11. By selecting the "Appearance" grouping, you can change the background color of the Chart Widget and format the text and symbol elements. Do so if you desire.

12. Save your work and click on the "Preview" button. Test out the Chart Widget by clicking on a census tract on the map.

Section 3: Adding Pages and Navigation

Section 3.1: Create a New Page

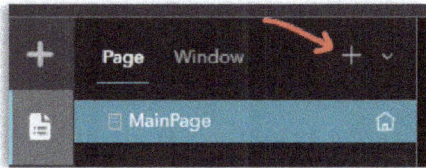

1. On the toolbar on the far-left side, select "Page" from the toolbar. On the "Page" tab, click the "+" sign in the upper right corner.

2. When the "Add page" window opens, select the "Scrolling page" tab, and then choose the "Blank scrolling" template. You have just added a second page to your Experience Builder site. In the Page pane on the far-right side, toggle on both the header and the footer. If you save and preview your site, you will find out that there is no way to navigate from one page to the other. Let's change that

Section 3.2: Add a Button Widget for Navigation and Publish

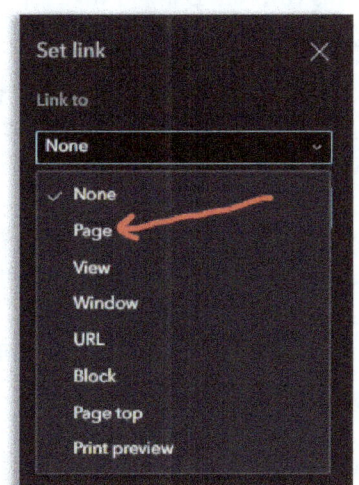

1. Since you are currently on the second page, let's add a Button Widget to the header that will switch from this page to the first when clicked. Begin by selecting "Edit header" at the top of the page. On the toolbar on the far-left side of the page, select the Insert Widget option. Search for the "Button" widget. Drag the Button Widget to the far-right side of the header. The "Quick style" window opens. This is an optional method of formatting; close it for this tutorial. In the Button pane, click on "Set link." The "Set link" window opens. From the "Link to" drop-down menu, choose "Page."

2. From the "Select a page" drop-down menu, choose the title of the first page of your site. Click "OK." In the Button pane, change the "Text" of the button to read "Page 1." By toggling "Advanced," you can format the button if you desire.

183

Additionally, by accessing the "Style" tab of the Button pane, you will see the familiar tools to adjust the size and position of the widget. It is recommended that positions are based on pixels rather than percentages.

Section 3. 2 Task (a): Add a second Button Widget to the header. This will be the button that navigates to your second page. Title it "Page 2." When you set the link for it, make sure to choose the name of the second page. Using the "Advanced" options in the Button pane, format both buttons.

3. Save your work and click on the "Preview" button. Test out the buttons' ability to navigate between pages.

Section 3.2 Task (b): Now that you have a second page, you need to populate it with content using the skills presented in this tutorial. On the second page, include a Map Widget that has the same webmap as the first page. You will also add widgets that you have not yet explored. Choose three widgets from the following list, add them to your second page, configure them, format them, and test them: "Print," "Near Me," "Business Analyst," "Basemap Gallery," "Measurement," "Table."

Should you need additional resources, consider referring to the ESRI documentation for widgets: https://doc.arcgis.com/en/experience-builder/latest/configure-widgets/widgets-overview.htm

Section 3.2 Task (c): Edit the footer in a way similar to that of the header near the beginning of this tutorial. Using a Text Widget, provide a summary of your site to the user.

4. The final thing to do is to publish your site. Once you are finished editing the site, in the upper-right hand corner, click on the "Publish" button.

5. Return to the content section of your ArcGIS Online page. Select your Experience Builder file. From the details page, click "View."

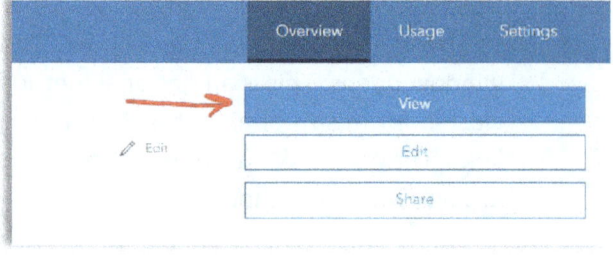

Section 3.2 Task (d): Copy the URL of your site and submit it.

ENGAGE WITH DR. ESRA OZDENEROL

Dr. Esra Ozdenerol, a Dunavant University Professor of Geographic Information Science, directs the Graduate Certificate in GIS at the University of Memphis and is an Adjunct Professor of Preventive Medicine at the University of Tennessee Health Science Center, where she teaches GIS applications to health disparities. Her research centers on applying geospatial technologies to precision health and environmental studies, with a focus on environmental health issues such as social determinants of health, infectious and vector-borne diseases, cardiovascular and birth outcomes, cancer, and lead poisoning.

She has consulted for various public and international agencies. Her textbook Spatial Health Inequalities: Adapting GIS Tools and Data Analysis is recognized by institutions like AAG and widely adopted. Her book Gender Inequalities: GIS Approaches to Gender Analysis explores how spatial analysis addresses gendered inequalities and promotes social justice. Her latest book The Role of GIS in COVID-19 Management and Control examines GIS applications in managing COVID-19, showcasing global projects and innovative GIS tools for outbreak control and decision-making.

She has lectured and led workshops on GIS applications in health at both national and international levels, sharing her expertise with a wide range of audiences, including public health professionals, researchers, and policymakers. Her presentations cover topics such as spatial analysis of health data, mapping disease patterns, and using GIS tools to inform health interventions and policy decisions.

Dr. Esra Ozdenerol teaches drone image processing and FAA Part 107 pilot preparation, offering hands-on instruction and tutorials to help students master both drone operations and certification requirements.

In addition to her academic work, Dr. Ozdenerol owns a GIS consulting company specializing in GIS applications for health and providing training for health professionals.

Stay up-to-date with Dr. Esra Ozdenerol's latest books by visiting her Amazon author page: amazon.com/author/esraozdenerol. Connect with her on Instagram @Prof.Dr.EsraOz and on Facebook @Prof.Dr.Esra Oz. For information about her GIS courses, books, and other GIS consultancy inquiries, visit her official website: www.esraozdenerol.com.

www.ingramcontent.com/pod-product-compliance
Lightning Source LLC
Chambersburg PA
CBHW080736300426
44114CB00019B/2608